Legal Notice

For information on bulk purchases and licensing agreements, please email

support@SATPrepGet800.com

ISBN-13: 978-1-951619-03-9

This is the Solution Guide to the book "Topology for Beginners."

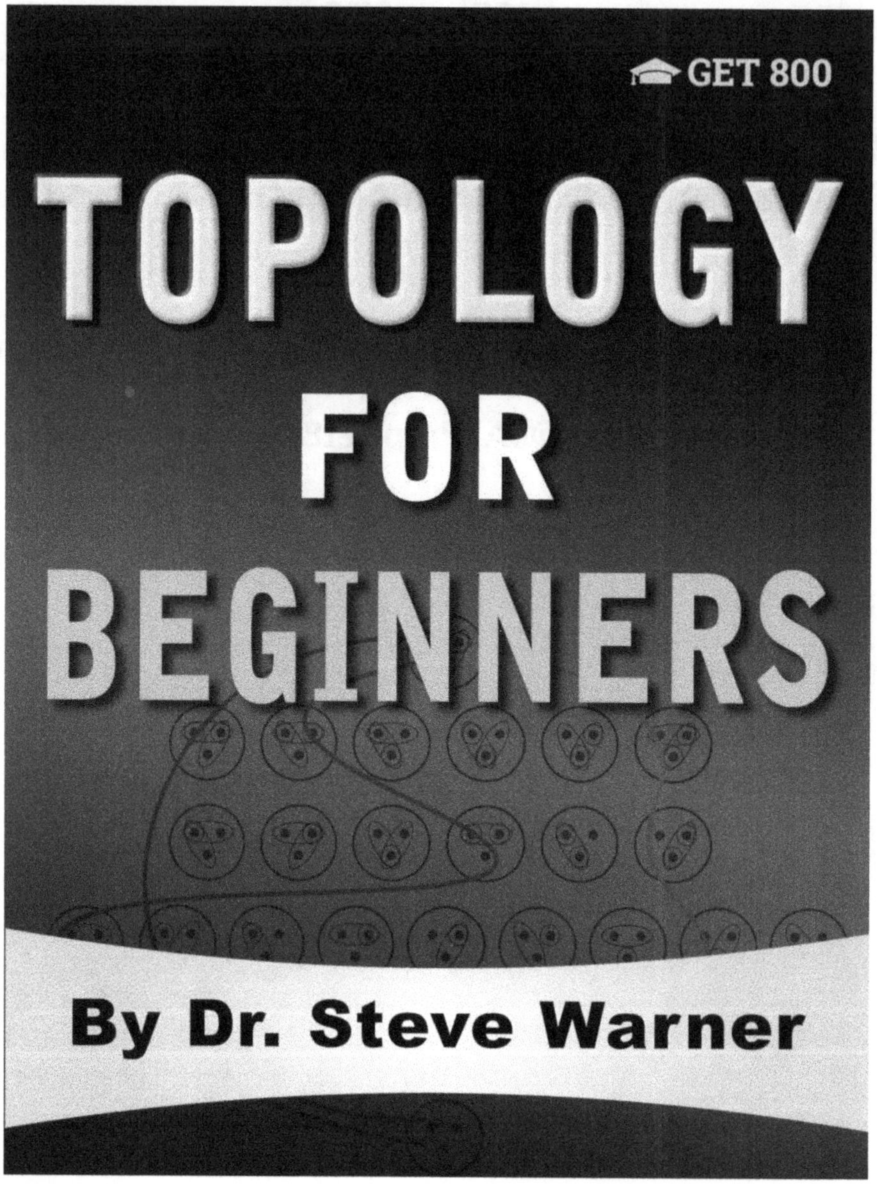

Also Available from Dr. Steve Warner

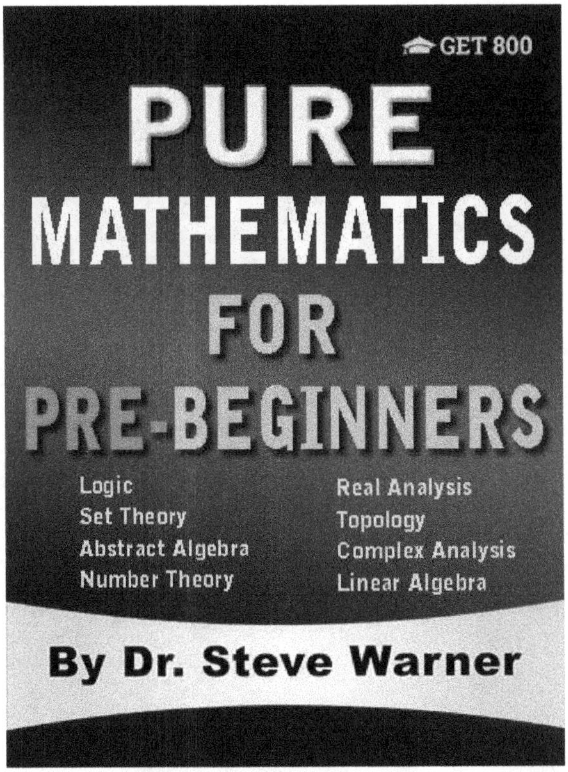

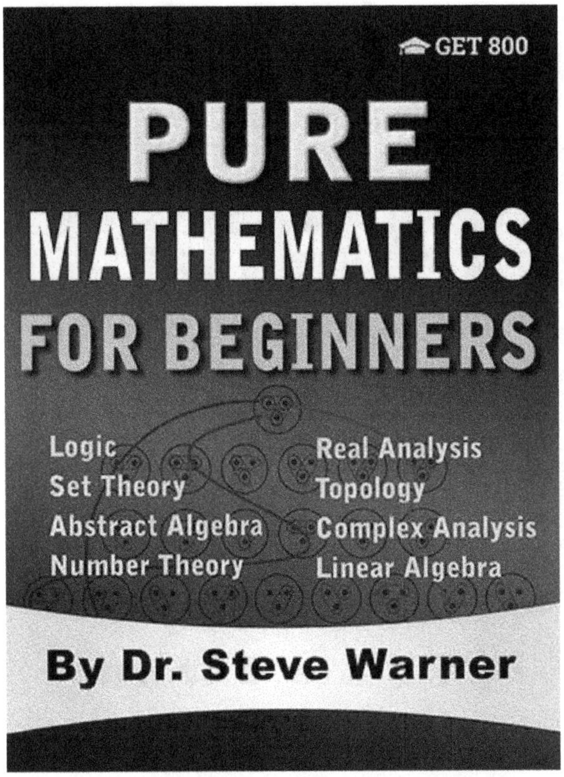

CONNECT WITH DR. STEVE WARNER

www.facebook.com/SATPrepGet800

www.youtube.com/TheSATMathPrep

www.twitter.com/SATPrepGet800

www.linkedin.com/in/DrSteveWarner

www.pinterest.com/SATPrepGet800

Also Available from Dr. Steve Warner

CONNECT WITH DR. STEVE WARNER

www.facebook.com/SATPrepGet800

www.youtube.com/TheSATMathPrep

www.twitter.com/SATPrepGet800

www.linkedin.com/in/DrSteveWarner

www.pinterest.com/SATPrepGet800

Topology for Beginners

Solution Guide

Dr. Steve Warner

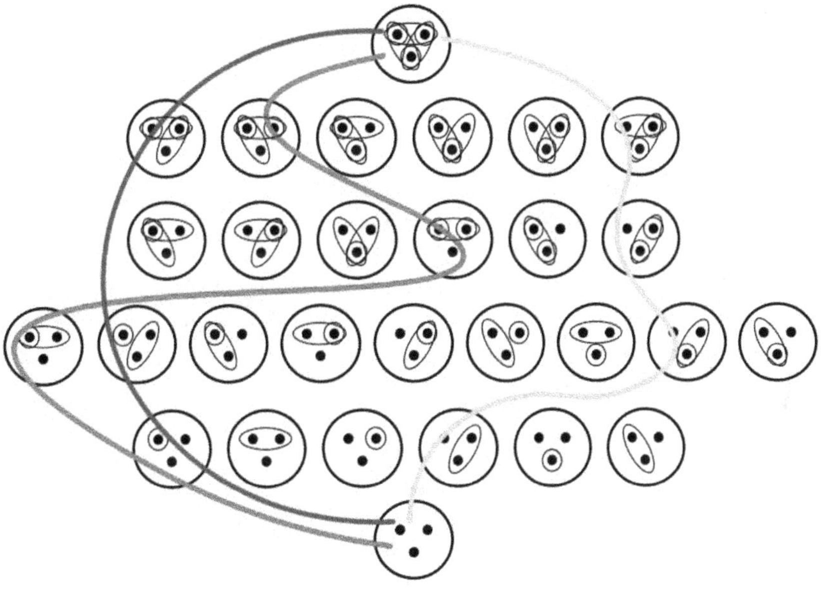

Table of Contents

Problem Set 1

LEVEL 1

1. Determine whether each of the following statements is true or false:

   (i) $7 \in \{7\}$

   (ii) $\delta \in \{\gamma, \delta, \epsilon\}$

   (iii) $-3 \in \{3\}$

   (iv) $3 \in \mathbb{Z}$

   (v) $-42 \in \mathbb{N}$

   (vi) $\frac{15}{17} \in \mathbb{Q}$

   (vii) $\emptyset \subseteq \{0,1,2\}$

   (viii) $\{\delta\} \subseteq \{\delta, \Delta\}$

   (ix) $\{x,y,z\} \subseteq \{x,y,z\}$

   (x) $\{a,b,\{c,d\}\} \subseteq \{a,b,c,d\}$

Solutions:

(i) $\{7\}$ has exactly 1 element, namely 7. So, $7 \in \{7\}$ is **true**.

(ii) $\{\gamma, \delta, \epsilon\}$ has exactly 3 elements, namely γ, δ, and ϵ. In particular, $\delta \in \{\gamma, \delta, \epsilon\}$ is **true**.

(iii) $\{3\}$ has exactly 1 element, namely 3. So, $-3 \notin \{3\}$. Therefore, $-3 \in \{3\}$ is **false**.

(iv) $\mathbb{Z} = \{\dots, -4, -3, -2, -1, 0, 1, 2, 3, 4, \dots\}$. In particular, $3 \in \mathbb{Z}$ is **true**.

(v) $\mathbb{N} = \{0, 1, 2, 3, \dots\}$. Therefore, $-42 \in \mathbb{N}$ is **false**.

(vi) Since $15, 17 \in \mathbb{Z}$ and $17 \neq 0$, $\frac{15}{17} \in \mathbb{Q}$ is **true**.

(vii) The empty set is a subset of every set. So, $\emptyset \subseteq \{0,1,2\}$ is **true**.

(viii) The only element of $\{\delta\}$ is δ. Since δ is also an element of $\{\delta, \Delta\}$, $\{\delta\} \subseteq \{\delta, \Delta\}$ is **true**.

(ix) Every set is a subset of itself. So, $\{x,y,z\} \subseteq \{x,y,z\}$ is **true**.

(x) $\{c,d\} \in \{a,b,\{c,d\}\}$, but $\{c,d\} \notin \{a,b,c,d\}$. So, $\{a,b,\{c,d\}\} \subseteq \{a,b,c,d\}$ is **false**.

2. Determine the cardinality of each of the following sets:

   (i) $\{\square, \Delta, \Gamma\}$

   (ii) $\{0,1,2,3,4,5\}$

   (iii) $\{1, 2, \dots, 37\}$

   (iv) $\left\{\frac{1}{2}, \frac{1}{3}, \dots, \frac{1}{11}\right\}$

Solutions:

(i) $|\{\square, \Delta, \Gamma\}| = \mathbf{3}$.

(ii) $\{0, 1, 2, 3, 4, 5\} = \mathbf{6}$.

(iii) $|\{1, 2, \dots, 37\}| = \mathbf{37}$.

(iv) $\left|\left\{\frac{1}{2}, \frac{1}{3}, \dots, \frac{1}{11}\right\}\right| = \mathbf{10}$.

3. Provide a single example of a set A with the following properties: (i) $A \subset \mathbb{C}$ (A is a *proper* subset of $\mathbb{C}$); (ii) A is infinite; (iii) A contains real numbers; and (iv) A contains complex numbers that are not real.

Solution: One example of such a set A is $A = \{a + bi \mid a \in \mathbb{R} \wedge (b = 0 \vee b = 1)\}$.

LEVEL 2

4. Compute the power set of each of the following sets:

(i) $\emptyset$

(ii) $\{5\}$

(iii) $\{m, n\}$

(iv) $\{\emptyset, \{\emptyset\}\}$

(v) $\{\{\emptyset\}\}$

Solutions:

(i) $\mathcal{P}(\emptyset) = \{\emptyset\}$

(ii) $\mathcal{P}(\{5\}) = \{\emptyset, \{5\}\}$

(iii) $\mathcal{P}(\{m, n\}) = \{\emptyset, \{m\}, \{n\}, \{m, n\}\}$

(iv) $\mathcal{P}(\{\emptyset, \{\emptyset\}\}) = \{\emptyset, \{\emptyset\}, \{\{\emptyset\}\}, \{\emptyset, \{\emptyset\}\}\}$

(v) $\mathcal{P}(\{\{\emptyset\}\}) = \{\emptyset, \{\{\emptyset\}\}\}$

5. Determine whether each of the following statements is true or false:

   (i) $3 \in \emptyset$

   (ii) $\emptyset \in \{a, b\}$

   (iii) $\emptyset \in \emptyset$

   (iv) $\emptyset \in \{\emptyset, \{\emptyset\}\}$

   (v) $\{\emptyset\} \in \emptyset$

   (vi) $\{\emptyset\} \in \{\emptyset\}$

   (vii) $\emptyset \subseteq \emptyset$

   (viii) $\emptyset \subseteq \{\emptyset\}$

   (ix) $\{\emptyset\} \subseteq \emptyset$

   (x) $\{\emptyset\} \subseteq \{\emptyset\}$

   (xi) $\mathbb{Q} \subseteq \mathbb{C}$

   (xii) $3 \in \{2k \mid k = 1, 2, 3, 4, 5, 6\}$

Solutions:

(i) The empty set has no elements. So, $x \in \emptyset$ is false for any x. In particular, $3 \in \emptyset$ is **false**.

(ii) $\{a, b\}$ has exactly 2 elements, namely a and b. In particular, $\emptyset \notin \{a, b\}$. So, $\emptyset \in \{a, b\}$ is **false**.

(iii) The empty set has no elements. So, $x \in \emptyset$ is false for any x. In particular, $\emptyset \in \emptyset$ is **false**.

(iv) The set $\{\emptyset, \{\emptyset\}\}$ has exactly 2 elements, namely $\emptyset$ and $\{\emptyset\}$. In particular, $\emptyset \in \{\emptyset, \{\emptyset\}\}$ is **true**.

(v) The empty set has no elements. So, $x \in \emptyset$ is false for any x. In particular, $\{\emptyset\} \in \emptyset$ is **false**.

(vi) The set $\{\emptyset\}$ has 1 element, namely $\emptyset$. Since $\{\emptyset\} \neq \emptyset$, $\{\emptyset\} \in \{\emptyset\}$ is **false**.

(vii) The empty set is a subset of every set. So, $\emptyset \subseteq X$ is true for any X. In particular, $\emptyset \subseteq \emptyset$ is **true**. (This can also be done by using the fact that every set is a subset of itself.)

(viii) Again, (as in (vii)), $\emptyset \subseteq X$ is true for any X. In particular, $\emptyset \subseteq \{\emptyset\}$ is **true**.

(ix) The only subset of $\emptyset$ is $\emptyset$. So, $\{\emptyset\} \subseteq \emptyset$ is **false**.

(x) Every set is a subset of itself. So, $\{\emptyset\} \subseteq \{\emptyset\}$ is **true**.

(xi) Since $\mathbb{Q} \subseteq \mathbb{R}$ and $\mathbb{R} \subseteq \mathbb{C}$, by Theorem 1.14, $\mathbb{Q} \subseteq \mathbb{C}$ is **true**.

(xii) $\{2k \mid k = 1, 2, 3, 4, 5, 6\} = \{2, 4, 6, 8, 10, 12\}$. So, $3 \notin \{2k \mid k = 1, 2, 3, 4, 5, 6\}$. Therefore, $3 \in \{2k \mid k = 1, 2, 3, 4, 5, 6\}$ is **false**.

6. Determine the cardinality of each of the following sets:

   (i) $\{a, a, b, c, d, d, d\}$

   (ii) $\{\{1, 2\}, \{3,4,5\}\}$

   (iii) $\{5, 6, 7, \ldots, 2122, 2123\}$

Solutions:

(i) $\{a,a,b,c,d,d,d\} = \{a,b,c,d\}$. Therefore, $|\{a,a,b,c,d,d,d\}| = |\{a,b,c,d\}| = \mathbf{4}$.

(ii) $\{\{1,2\},\{3,4,5\}\}$ consists of the 2 elements $\{1,2\}$ and $\{3,4,5\}$. So, $|\{\{1,2\},\{3,4,5\}\}| = \mathbf{2}$.

(iii) $|\{5,6,7,\ldots,2122,2123\}| = 2123 - 5 + 1 = \mathbf{2119}$.

Note: For number (iii), we used the fence-post formula (see Notes 3 and 4 after Example 1.7).

LEVEL 3

7. Determine the cardinality of each of the following sets:

   (i) $\{\{\{a,b\}\}\}$

   (ii) $\{\{0,1\},0,\{0\},\{0,\{0,1,2\}\}\}$

   (iii) $\{a,\{a\},\{a,a\},\{a,a,a,a\},\{a,a,\{a\}\},\{a,\{a\},\{a\}\}\}$

Solutions:

(i) The only element of $\{\{\{a,b\}\}\}$ is $\{\{a,b\}\}$. So, $|\{\{\{a,b\}\}\}| = \mathbf{1}$.

(ii) The elements of $\{\{0,1\},0,\{0\},\{0,\{0,1,2\}\}\}$ are $\{0,1\}$, 0, $\{0\}$, and $\{0,\{0,1,2\}\}$. So, we see that $|\{\{0,1\},0,\{0\},\{0,\{0,1,2\}\}\}| = \mathbf{4}$.

(iii) We have:
$$\begin{aligned}\{a,\{a\},\{a,a\},\{a,a,a,a\},\{a,a,\{a\}\},\{a,\{a\},\{a\}\}\} &= \{a,\{a\},\{a\},\{a\},\{a,\{a\}\},\{a,\{a\}\}\} \\ &= \{a,\{a\},\{a,\{a\}\}\}.\end{aligned}$$

So, $|\{a,\{a\},\{a,a\},\{a,a,a,a\},\{a,a,\{a\}\},\{a,\{a\},\{a\}\}\}| = |\{a,\{a\},\{a,\{a\}\}\}| = \mathbf{3}$.

8. How many subsets does $\{a,b,c,d\}$ have? Draw a tree diagram for the subsets of $\{a,b,c,d\}$.

Solution: $|\{a,b,c,d\}| = 4$. Therefore, $\{a,b,c,d\}$ has $2^4 = \mathbf{16}$ subsets. We can also say that the size of the power set of $\{a,b,c,d\}$ is 16, that is, $|\mathcal{P}(\{a,b,c,d\})| = 16$. Here is a tree diagram.

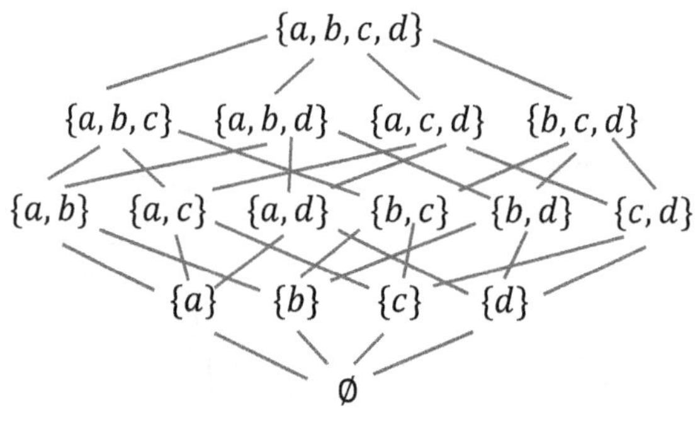

9. Let A, B, C, D, and E be sets such that $A \subseteq B, B \subseteq C, C \subseteq D$, and $D \subseteq E$. Prove that $A \subseteq E$.

Proof: Suppose that A, B, C, D, and E are sets such that $A \subseteq B$, $B \subseteq C$, $C \subseteq D$, and $D \subseteq E$. Since $A \subseteq B$ and $B \subseteq C$, by Theorem 1.14, we have $A \subseteq C$. Since $A \subseteq C$ and $C \subseteq D$, again by Theorem 1.14, we have $A \subseteq D$. Finally, since $A \subseteq D$ and $D \subseteq E$, once again by Theorem 1.14, we have $A \subseteq E$. □

LEVEL 4

10. A relation R is **reflexive** if $\forall x(xRx)$ and **symmetric** if $\forall x \forall y(xRy \to yRx)$. For example, the relation "=" is reflexive and symmetric because $\forall x(x = x)$ and $\forall x \forall y(x = y \to y = x)$. Show that $\subseteq$ is reflexive, but $\in$ is not. Then decide if each of $\subseteq$ and $\in$ is symmetric.

Solutions: (**$\subseteq$ is reflexive**) Let A be a set. By Theorem 2.1, A is a subset of itself. So, $A \subseteq A$ is true. Since A was arbitrary, $\forall x(A \subseteq A)$ is true. Therefore, $\subseteq$ is reflexive. □

($\in$ is **not** reflexive) Since the empty set has no elements, $\emptyset \notin \emptyset$. This **counterexample** shows that $\in$ is not reflexive.

($\subseteq$ is **not** symmetric) $\{1\} \subseteq \{1,2\}$, but $\{1,2\} \nsubseteq \{1\}$. This **counterexample** shows that $\subseteq$ is not symmetric.

($\in$ is **not** symmetric) $\emptyset \in \{\emptyset\}$, but $\{\emptyset\} \notin \emptyset$. This **counterexample** shows that $\in$ is not symmetric.

Note: A **conjecture** is an educated guess. In math, conjectures are made all the time based upon evidence from examples (but examples alone cannot be used to prove a conjecture). A logical argument is usually needed to prove a conjecture, whereas a single **counterexample** is used to disprove a conjecture. For example, $\emptyset \notin \emptyset$ is a counterexample to the conjecture "$\in$ is reflexive."

11. Determine whether each of the following statements is true or false:

(i) $c \in \{a, \{c\}\}$

(ii) $\{\Delta\} \in \{\delta, \Delta\}$

(iii) $\{1\} \in \{1, a, 2, b\}$

(iv) $\emptyset \in \{\{\emptyset\}\}$

(v) $\{\{\emptyset\}\} \in \emptyset$

Solutions:

(i) The set $\{a, \{c\}\}$ has exactly 2 elements, namely a and $\{c\}$. So, $c \in \{a, \{c\}\}$ is **false**.

(ii) The set $\{\delta, \Delta\}$ has exactly 2 elements, namely δ and Δ. So, $\{\Delta\} \in \{\delta, \Delta\}$ is **false**.

(iii) The set $\{1, a, 2, b\}$ has exactly 4 elements, namely $1, a, 2$, and b. So, $\{1\} \in \{1, a, 2, b\}$ is **false**.

(iv) The set $\{\{\emptyset\}\}$ has exactly 1 element, namely $\{\emptyset\}$. Since $\emptyset$ is not equal to $\{\emptyset\}$, $\emptyset \in \{\{\emptyset\}\}$ is **false**.

(v) The empty set has no elements. So, $x \in \emptyset$ is false for any x. In particular, $\{\{\emptyset\}\} \in \emptyset$ is **false**.

12. We say that a set A is **transitive** if $\forall x(x \in A \to x \subseteq A)$. Determine if each of the following sets is transitive:

(i) $\emptyset$

(ii) $\{\emptyset\}$

(iii) $\{\{\emptyset\}\}$

(iv) $\{\emptyset, \{\emptyset\}\}$

(v) $\{\emptyset, \{\emptyset\}, \{\{\emptyset\}\}\}$

(vi) $\{\{\emptyset\}, \{\emptyset, \{\emptyset\}\}\}$

Solutions:

(i) Since $\emptyset$ has no elements, $\emptyset$ **is transitive**. (The statement "$x \in \emptyset \to x \subseteq \emptyset$" is true simply because "$x \in \emptyset$" is always false. In this case, we say that the statement is vacuously true.)

(ii) The only element of $\{\emptyset\}$ is $\emptyset$, and $\emptyset \subseteq \{\emptyset\}$ is true. So, $\{\emptyset\}$ **is transitive**.

(iii) $\{\emptyset\} \in \{\{\emptyset\}\}$ and $\emptyset \in \{\emptyset\}$, but $\emptyset \notin \{\{\emptyset\}\}$. So, $\{\{\emptyset\}\}$ **is not transitive**.

(iv) $\{\emptyset, \{\emptyset\}\}$ has 2 elements, namely $\emptyset$ and $\{\emptyset\}$. Both sets are subsets of $\{\emptyset, \{\emptyset\}\}$. It follows that $\{\emptyset, \{\emptyset\}\}$ **is transitive**.

(v) $\{\emptyset, \{\emptyset\}, \{\{\emptyset\}\}\}$ has 3 elements, namely $\emptyset$, $\{\emptyset\}$, and $\{\{\emptyset\}\}$. All three of these sets are subsets of $\{\emptyset, \{\emptyset\}, \{\{\emptyset\}\}\}$. It follows that $\{\emptyset, \{\emptyset\}, \{\{\emptyset\}\}\}$ **is transitive**.

(vi) $\{\emptyset\} \in \{\{\emptyset\}, \{\emptyset, \{\emptyset\}\}\}$ and $\emptyset \in \{\emptyset\}$, but $\emptyset \notin \{\{\emptyset\}, \{\emptyset, \{\emptyset\}\}\}$. So, $\{\{\emptyset\}, \{\emptyset, \{\emptyset\}\}\}$ **is not transitive**.

LEVEL 5

13. Let A and B be sets with $A \subseteq B$. Prove that $\mathcal{P}(A) \subseteq \mathcal{P}(B)$.

Proof: Let A and B be sets with $A \subseteq B$ and let $X \in \mathcal{P}(A)$. Then $X \subseteq A$ Since $X \subseteq A$ and $A \subseteq B$, by Theorem 1.14, $X \subseteq B$. So, $X \in \mathcal{P}(B)$. Since $X \in \mathcal{P}(A)$ was arbitrary, $\forall X(X \in \mathcal{P}(A) \to X \in \mathcal{P}(B))$. Therefore, $\mathcal{P}(A) \subseteq \mathcal{P}(B)$. □

14. Prove that if A is a transitive set, then $\mathcal{P}(A)$ is also a transitive set (see Problem 12 above for the definition of a transitive set).

Proof: Let A be a transitive set, let $x \in \mathcal{P}(A)$, and let $y \in x$. Since $x \in \mathcal{P}(A)$, $x \subseteq A$. Since $y \in x$ and $x \subseteq A$, $y \in A$. Since A is transitive and $y \in A$, $y \subseteq A$. So, $y \in \mathcal{P}(A)$. Since $y \in x$ was arbitrary, $\forall y(y \in x \to y \in \mathcal{P}(A))$. Therefore, $x \subseteq \mathcal{P}(A)$. Since x was arbitrary, $\forall x(x \in \mathcal{P}(A) \to x \subseteq \mathcal{P}(A))$. Thus, $\mathcal{P}(A)$ is transitive. □

15. Let $P(x)$ be the property $x \notin x$. Prove that $\{x | P(x)\}$ cannot be a set.

Solution: Suppose toward contradiction that $A = \{x \mid x \notin x\}$ is a set. Then $A \in A$ if and only if $A \notin A$. So, $p \leftrightarrow \neg p$ is true, where p is the statement $A \in A$. However, $p \leftrightarrow \neg p$ is always false. This is a contradiction. So, A is not a set. □

Notes: (1) This is our first **proof by contradiction**. A proof by contradiction works as follows:

1. We assume the negation of what we are trying to prove.
2. We use a logically sound argument to derive a statement which is false.
3. Since the argument is logically sound, the only possible error is our original assumption. Therefore, the negation of our original assumption must be true.

In this problem we are trying to prove that $A = \{x \mid x \notin x\}$ **is not** a set. The negation of this statement is that $A = \{x \mid x \notin x\}$ **is** a set. We then use only the definition of A to get the false statement $A \in A \leftrightarrow \neg A \in A$. Since the argument was logically valid, our initial assumption must have been incorrect, and therefore, A is not a set.

(2) The contradiction that occurs here is known as **Russell's Paradox**. This contradiction shows that we need to be careful about how we define a set. A naïve definition would be that a set is any object that has the form $\{x|P(x)\}$, where $P(x)$ is an arbitrary property (by property, we mean a **first-order property**—this is a property defined using the connectives $\wedge$, $\vee$, $\rightarrow$, and $\leftrightarrow$, the quantifiers $\forall$ and $\exists$, and the relations $=$ and $\in$). As we see in this problem, that "definition" of a set leads to a contradiction. Instead, we call $\{x|P(x)\}$ a **class**. Every set is a class, but not every class is a set. A class that is not a set is called a **proper class**. For example, $\{x|x \notin x\}$ is a proper class.

16. Let $A = \{a, b, c, d\}$, $B = \{X \mid X \subseteq A \wedge d \notin X\}$, and $C = \{X \mid X \subseteq A \wedge d \in X\}$. Show that there is a natural one-to-one correspondence between the elements of B and the elements of C. Then generalize this result to a set with $n + 1$ elements for $n > 0$.

Solution: We define the one-to-one correspondence as follows: If $Y \in B$, then Y is a subset of A that does not contain d. Let Y_d be the set that contains the same elements as Y, but with d thrown in. Then the correspondence $Y \rightarrow Y_d$ is a one-to-one correspondence. We can see this correspondence in the table below.

Elements of B	Elements of C
$\emptyset$	$\{d\}$
$\{a\}$	$\{a, d\}$
$\{b\}$	$\{b, d\}$
$\{c\}$	$\{c, d\}$
$\{a, b\}$	$\{a, b, d\}$
$\{a, c\}$	$\{a, c, d\}$
$\{b, c\}$	$\{b, c, d\}$
$\{a, b, c\}$	$\{a, b, c, d\}$

For the general result, we start with a set A with $n+1$ elements, and we let d be some element from A. Define B and C the same way as before: $B = \{X \mid X \subseteq A \wedge d \notin X\}$, and $C = \{X \mid X \subseteq A \wedge d \in X\}$. Also, as before, if $Y \in B$, then Y is a subset of A that does not contain d. Let Y_d be the set that contains the same elements as Y, but with d thrown in. Then the correspondence $Y \to Y_d$ is a one-to-one correspondence.

Notes: (1) B consists of the subsets of A that do not contain the element d, while C consists of the subsets of A that do contain d.

(2) Observe that in the case where $A = \{a, b, c, d\}$, B and C each have $8 = 2^3$ elements. Also, there is no overlap between B and C (they have no elements in common). So, we have a total of $8 + 8 = 16$ elements. Since there are exactly $2^4 = 16$ subsets of A, we see that we have listed every subset of A.

(3) We could also do the computation in Note 2 as follows: $2^3 + 2^3 = 2 \cdot 2^3 = 2^1 \cdot 2^3 = 2^{1+3} = 2^4$. It's nice to see the computation this way because it mimics the computation we will do in the more general case. In case your algebra skills are not that strong, here is an explanation of each step:

Adding the same thing to itself is equivalent to multiplying that thing by 2. For example, 1 apple plus 1 apple is 2 apples. Similarly, $1x + 1x = 2x$. This could be written more briefly as $x + x = 2x$. Replacing x by 2^3 gives us $2^3 + 2^3 = 2 \cdot 2^3$ (the first equality in the computation above).

Next, by definition, $x^1 = x$. So, $2^1 = 2$. Therefore, we can rewrite $2 \cdot 2^3$ as $2^1 \cdot 2^3$.

Now, 2^3 means to multiply 2 by itself 3 times. So, $2^3 = 2 \cdot 2 \cdot 2$. Thus, $2^1 \cdot 2^3 = 2 \cdot 2 \cdot 2 \cdot 2 = 2^4$. This leads to the rule of exponents which says that if you multiply two expressions with the same base, you can add the exponents. So, $2^1 \cdot 2^3 = 2^{1+3} = 2^4$.

(4) In the more general case, B and C each have 2^n elements. The reason for this is that A has $n+1$ elements. When we remove the element d from A, the resulting set has n elements, and therefore, 2^n subsets. B consists of precisely the subsets of this new set (A with d removed), and so, B has exactly 2^n elements. The one-to-one correspondence $Y \to Y_d$ shows that C has the same number of elements as B. Therefore, C also has 2^n elements.

(5) In the general case, there is still no overlap between B and C. It follows that the total number of elements when we combine B and C is $2^n + 2^n = 2 \cdot 2^n = 2^1 \cdot 2^n = 2^{1+n} = 2^{n+1}$. See Note 3 above for an explanation as to how all this algebra works.

(6) By a **one-to-one correspondence** between the elements of B and the elements of C, we mean a pairing where we match each element of B with exactly one element of C so that each element of C is matched with exactly one element of B. The table given in the solution above provides a nice example of such a pairing.

(7) In the case where $A = \{a, b, c, d\}$, B consists of all the subsets of $\{a, b, c\}$. In other words, $B = \{X \mid X \subseteq \{a, b, c\}\} = \mathcal{P}(\{a, b, c\})$.

A description of C is a bit more complicated. It consists of the subsets of $\{a, b, c\}$ with d thrown into them. We could write this as $C = \{X \cup \{d\} \mid X \subseteq \{a, b, c\}\}$.

(5) In the general case, we can write $K = A \setminus \{d\}$ (this is the set consisting of all the elements of A, except d). We then have $B = \{X \mid X \subseteq K\} = \mathcal{P}(K)$ and $C = \{X \cup \{d\} \mid X \subseteq K\} = \mathcal{P}(A) \setminus \mathcal{P}(K)$.

(6) The symbols "∪" for **union** and "\" for **set difference** will be defined formally in Lesson 2.

Problem Set 2

LEVEL 1

1. Let $A = \{a, b, \Delta, \delta\}$ and $B = \{b, c, \delta, \gamma\}$. Determine each of the following:

   (i) $A \cup B$

   (ii) $A \cap B$

   (iii) $A \setminus B$

   (iv) $B \setminus A$

   (v) $A \,\Delta\, B$

Solutions:

(i) $A \cup B = \{a, b, c, \Delta, \delta, \gamma\}$.

(ii) $A \cap B = \{b, \delta\}$.

(iii) $A \setminus B = \{a, \Delta\}$

(iv) $B \setminus A = \{c, \gamma\}$

(v) $A \,\Delta\, B = \{a, \Delta\} \cup \{c, \gamma\} = \{a, c, \Delta, \gamma\}$

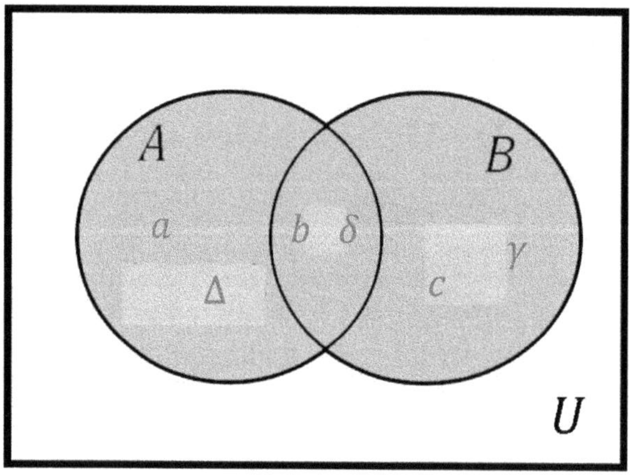

2. Draw Venn diagrams for $(A \setminus B) \setminus C$ and $A \setminus (B \setminus C)$. Are these two sets equal for all sets A, B, and C? If so, prove it. If not, provide a counterexample.

Solution:

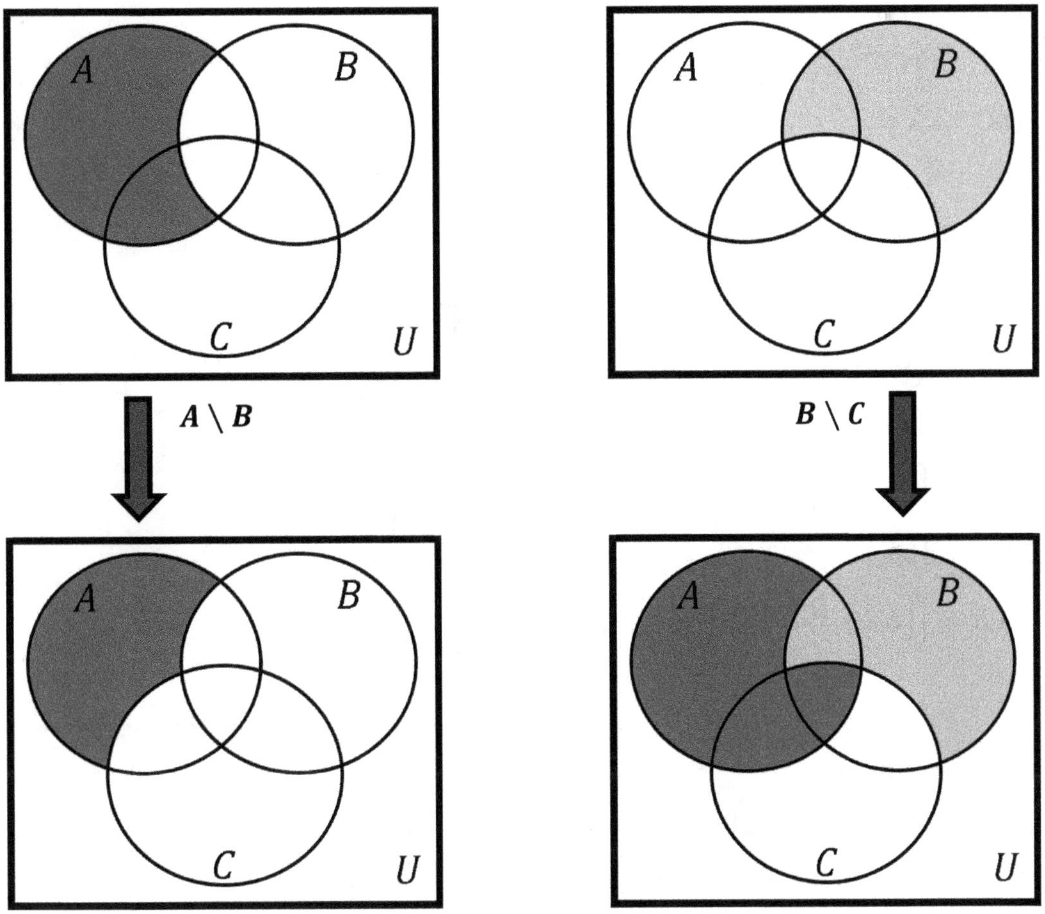

$(A \setminus B) \setminus C$ $\qquad$ $A \setminus (B \setminus C)$

From the Venn diagrams, it looks like $(A \setminus B) \setminus C \subseteq A \setminus (B \setminus C)$, but $(A \setminus B) \setminus C \neq A \setminus (B \setminus C)$.

Let's come up with a counterexample. Let $A = \{1, 2\}$, $B = \{1,3\}$, and $C = \{1, 4\}$. Then we have $(A \setminus B) \setminus C = \{2\} \setminus \{1, 4\} = \{2\}$ and $A \setminus (B \setminus C) = \{1, 2\} \setminus \{3\} = \{1, 2\}$.

We see that $(A \setminus B) \setminus C \neq A \setminus (B \setminus C)$.

Note: Although it was not asked in the question, let's prove that $(A \setminus B) \setminus C \subseteq A \setminus (B \setminus C)$. Let $x \in (A \setminus B) \setminus C$. Then $x \in A \setminus B$ and $x \notin C$. Since $x \in A \setminus B$, $x \in A$ and $x \notin B$. In particular, $x \in A$. Since $x \notin B$, $x \notin B \setminus C$ (because if $x \in B \setminus C$, then $x \in B$). So, we have $x \in A$ and $x \notin B \setminus C$. Therefore, $x \in A \setminus (B \setminus C)$. Since $x \in (A \setminus B) \setminus C$ was arbitrary, $(A \setminus B) \setminus C \subseteq A \setminus (B \setminus C)$. □

LEVEL 2

3. Let $A = \{\emptyset, \{\emptyset, \{\emptyset\}\}\}$ and $B = \{\emptyset, \{\emptyset\}\}$. Compute each of the following:

    (i) $A \cup B$

    (ii) $A \cap B$

    (iii) $A \setminus B$

    (iv) $B \setminus A$

    (v) $A \,\Delta\, B$

Solutions:

(i) $A \cup B = \{\emptyset, \{\emptyset\}, \{\emptyset, \{\emptyset\}\}\}$.

(ii) $A \cap B = \{\emptyset\}$.

(iii) $A \setminus B = \{\{\emptyset, \{\emptyset\}\}\}$

(iv) $B \setminus A = \{\{\emptyset\}\}$

(v) $A \,\Delta\, B = \{\{\emptyset, \{\emptyset\}\}\} \cup \{\{\emptyset\}\} = \{\{\emptyset\}, \{\emptyset, \{\emptyset\}\}\}$

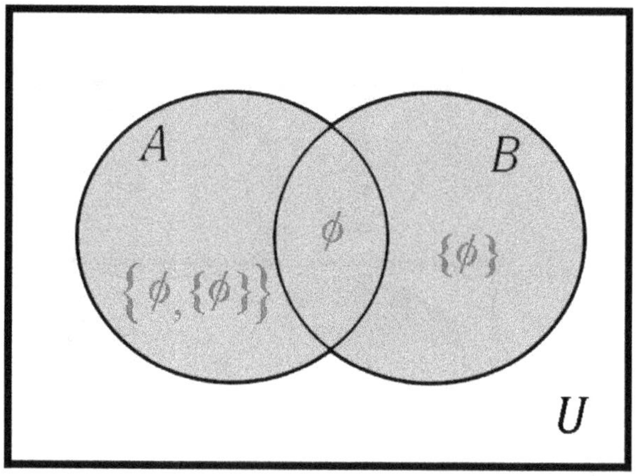

4. Prove the following:

    (i) The operation of forming unions is commutative.

    (ii) The operation of forming intersections is commutative.

    (iii) The operation of forming intersections is associative.

Proofs:

(i) Let A and B be sets. Then $x \in A \cup B$ if and only if $x \in A$ or $x \in B$ if and only if $x \in B$ or $x \in A$ if and only if $x \in B \cup A$. Since x was arbitrary, we have shown $\forall x(x \in A \cup B \leftrightarrow x \in B \cup A)$. Therefore, $A \cup B = B \cup A$. So, the operation of forming unions is commutative. □

(ii) Let A and B be sets. Then $x \in A \cap B$ if and only if $x \in A$ and $x \in B$ if and only if $x \in B$ and $x \in A$ if and only if $x \in B \cap A$. Since x was arbitrary, we have $\forall x(x \in A \cap B \leftrightarrow x \in B \cap A)$. Therefore, $A \cap B = B \cap A$. So, the operation of forming intersections is commutative. □

(iii) Let A, B, and C be sets. Then $x \in (A \cap B) \cap C$ if and only if $x \in A \cap B$ and $x \in C$ if and only if $x \in A$, $x \in B$ and $x \in C$ if and only if $x \in A$ and $x \in B \cap C$ if and only if $x \in A \cap (B \cap C)$. Since x was arbitrary, we have shown $\forall x\big(x \in (A \cap B) \cap C \leftrightarrow x \in A \cap (B \cap C)\big)$.

Therefore, we have shown that $(A \cap B) \cap C = A \cap (B \cap C)$. So, the operation of forming intersections is associative. □

LEVEL 3

5. Prove or provide a counterexample:
   (i) Every pairwise disjoint set of sets is disjoint.
   (ii) Every disjoint set of sets is pairwise disjoint.

Solutions:

(i) This is **false**. Let $A = \{1\}$ and let $\boldsymbol{X} = \{A\}$. $\boldsymbol{X}$ is pairwise disjoint, but $\bigcap \boldsymbol{X} = A = \{1\} \neq \emptyset$.

However, the following slightly modified statement is **true**: "Every pairwise disjoint set of sets consisting of at least two sets is disjoint."

Let $\boldsymbol{X}$ be a pairwise disjoint set of sets with at least two sets, say $A, B \in \boldsymbol{X}$. Suppose towards contradiction that $x \in \bigcap \boldsymbol{X}$. Then $x \in A$ and $x \in B$. So, $x \in A \cap B$. But $A \cap B = \emptyset$ because $\boldsymbol{X}$ is pairwise disjoint. This contradiction shows that the statement $x \in \bigcap \boldsymbol{X}$ is false. Therefore, $\boldsymbol{X}$ is disjoint. □

(ii) This is **false**. Let $A = \{0,1\}$, $B = \{1,2\}$, $C = \{0,2\}$, and $\boldsymbol{X} = \{A, B, C\}$. Then $\boldsymbol{X}$ is disjoint because $\bigcap \boldsymbol{X} = A \cap B \cap C = \{0,1\} \cap \{1,2\} \cap \{0,2\} = \{1\} \cap \{0,2\} = \emptyset$. However, $\boldsymbol{X}$ is **not** pairwise disjoint because $A \cap B = \{0,1\} \cap \{1,2\} = \{1\} \neq \emptyset$.

6. Let A and B be sets. Prove that $A \cap B \subseteq A$.

Proof: Suppose that A and B are sets and let $x \in A \cap B$. Then $x \in A$ and $x \in B$. In particular, $x \in A$. Since x was an arbitrary element of A, we have shown that every element of $A \cap B$ is an element of A. That is, $\forall x(x \in A \cap B \rightarrow x \in A)$ is true. Therefore, $A \cap B \subseteq A$. □

LEVEL 4

7. Prove that $B \subseteq A$ if and only if $A \cap B = B$.

Proof: Suppose that $B \subseteq A$. By part (ii) of Problem 4 and Problem 6, $A \cap B = B \cap A \subseteq B$. Let $x \in B$. Since $B \subseteq A$, $x \in A$. Therefore, $x \in A$ and $x \in B$. So, $x \in A \cap B$. Since x was an arbitrary element of B, we have shown that every element of B is an element of $A \cap B$. That is, $\forall x(x \in B \rightarrow x \in A \cap B)$. Therefore, $B \subseteq A \cap B$. Since $A \cap B \subseteq B$ and $B \subseteq A \cap B$, it follows that $A \cap B = B$.

Now, suppose that $A \cap B = B$ and let $x \in B$. Then $x \in A \cap B$. So, $x \in A$ and $x \in B$. In particular, $x \in A$. Since x was an arbitrary element of B, we have shown that every element of B is an element of A. That is, $\forall x(x \in B \rightarrow x \in A)$. Therefore, $B \subseteq A$. □

8. Let A, B, and C be sets. Prove each of the following:
   (i) $A \cap (B \cup C) = (A \cap B) \cup (A \cap C)$.
   (ii) $A \cup (B \cap C) = (A \cup B) \cap (A \cup C)$.
   (iii) $C \setminus (A \cup B) = (C \setminus A) \cap (C \setminus B)$.
   (iv) $C \setminus (A \cap B) = (C \setminus A) \cup (C \setminus B)$.

Proofs:

(i) $x \in A \cap (B \cup C) \Leftrightarrow x \in A$ and $x \in B \cup C \Leftrightarrow x \in A$ and either $x \in B$ or $x \in C \Leftrightarrow x \in A$ and $x \in B$ or $x \in A$ and $x \in C \Leftrightarrow x \in A \cap B$ or $x \in A \cap C \Leftrightarrow x \in (A \cap B) \cup (A \cap C)$. □

(ii) $x \in A \cup (B \cap C) \Leftrightarrow x \in A$ or $x \in B \cap C \Leftrightarrow$ either $x \in A$ or we have both $x \in B$ and $x \in C \Leftrightarrow$ we have both $x \in A$ or $x \in B$ and $x \in A$ or $x \in C \Leftrightarrow x \in A \cup B$ and $x \in A \cup C \Leftrightarrow$ $x \in (A \cup B) \cap (A \cup C)$. □

(iii) $x \in C \setminus (A \cup B) \Leftrightarrow x \in C$ and $x \notin A \cup B \Leftrightarrow x \in C$ and $x \notin A$ and $x \notin B \Leftrightarrow x \in C$ and $x \notin A$ and $x \in C$ and $x \notin B \Leftrightarrow x \in C \setminus A$ and $x \in C \setminus B \Leftrightarrow x \in (C \setminus A) \cap (C \setminus B)$. □

(iv) $x \in C \setminus (A \cap B) \Leftrightarrow x \in C$ and $x \notin A \cap B \Leftrightarrow x \in C$ and $x \notin A$ or $x \notin B \Leftrightarrow x \in C$ and $x \notin A$ or $x \in C$ and $x \notin B \Leftrightarrow x \in C \setminus A$ or $x \in C \setminus B \Leftrightarrow x \in (C \setminus A) \cup (C \setminus B)$. □

Notes: Let's let p, q, and r be the statements $x \in A$, $x \in B$, and $x \in C$, respectively.

(1) In (i) above, the statement "$x \in A$ and either $x \in B$ or $x \in C$" can be written $p \wedge (q \vee r)$. It can be shown that this is equivalent to $(p \wedge q) \vee (p \wedge r)$. In words, this is the statement "$x \in A$ and $x \in B$ or $x \in A$ and $x \in C$." Here it needs to be understood that the word "and" takes precedence over the word "or."

Similarly, we can use the logical equivalence $p \vee (q \wedge r) \equiv (p \vee q) \wedge (p \vee r)$ to help understand the proof of (ii).

(2) The equivalences $p \wedge (q \vee r) \equiv (p \wedge q) \vee (p \wedge r)$ and $p \vee (q \wedge r) \equiv (p \vee q) \wedge (p \vee r)$ are known as the **distributive laws**.

The rules $A \cap (B \cup C) = (A \cap B) \cup (A \cap C)$ and $A \cup (B \cap C) = (A \cup B) \cap (A \cup C)$ are also known as the **distributive laws**.

(3) To clarify (iii) and (iv), note that $\neg(p \vee q) \equiv \neg p \wedge \neg q$ and $\neg(p \wedge q) \equiv \neg p \vee \neg q$ (these equivalences can be easily checked). These two equivalences are known as **De Morgan's laws**. For (iii), we can use the logical equivalence $\neg(p \vee q) \equiv \neg p \wedge \neg q$ with p the statement $x \in A$ and q the statement $x \in B$ to get

$$x \notin A \cup B \equiv \neg x \in A \cup B \equiv \neg(x \in A \vee x \in B) \equiv \neg(p \vee q) \equiv \neg p \wedge \neg q \text{ (by De Morgan's law)}$$
$$\equiv \neg x \in A \wedge \neg x \in B \equiv x \notin A \wedge x \notin B.$$

So, the statement "$x \in C$ and $x \notin A \cup B$" is equivalent to $x \in C \wedge x \notin A \wedge x \notin B$.

Similarly, we can use the logical equivalence $\neg(p \wedge q) \equiv \neg p \vee \neg q$ to see that the statement "$x \in C$ and $x \notin A \cap B$" is equivalent to "$x \in C$ and $x \notin A$ or $x \notin B$."

(4) The rules $C \setminus (A \cup B) = (C \setminus A) \cap (C \setminus B)$ and $C \setminus (A \cap B) = (C \setminus A) \cup (C \setminus B)$ are also known as **De Morgan's laws**.

LEVEL 5

9. Let $\boldsymbol{X}$ be a nonempty set of sets. Prove the following:

    (i) For all $A \in \boldsymbol{X}$, $A \subseteq \bigcup \boldsymbol{X}$.

    (ii) For all $A \in \boldsymbol{X}$, $\bigcap \boldsymbol{X} \subseteq A$.

Proofs:

(i) Let $\boldsymbol{X}$ be a nonempty set of sets, let $A \in \boldsymbol{X}$, and let $x \in A$. Then there is $B \in \boldsymbol{X}$ such that $x \in B$ (namely A). So, $x \in \bigcup \boldsymbol{X}$. Since x was an arbitrary element of A, we have shown that $A \subseteq \bigcup \boldsymbol{X}$. Since A was an arbitrary element of $\boldsymbol{X}$, we have shown that for all $A \in \boldsymbol{X}$, we have $A \subseteq \bigcup \boldsymbol{X}$. □

(ii) Let $\boldsymbol{X}$ be a nonempty set of sets, let $A \in \boldsymbol{X}$, and let $x \in \bigcap \boldsymbol{X}$. Then for every $B \in \boldsymbol{X}$, we have $x \in B$. In particular, $x \in A$ (because $A \in \boldsymbol{X}$). Since x was an arbitrary element of $\bigcap \boldsymbol{X}$, we have shown that $\bigcap \boldsymbol{X} \subseteq A$. Since A was an arbitrary element of $\boldsymbol{X}$, we have shown that for all $A \in \boldsymbol{X}$, we have $\bigcap \boldsymbol{X} \subseteq A$. □

10. Let A be a set and let $\boldsymbol{X}$ be a nonempty set of sets. Prove each of the following:

    (i) $A \cap \bigcup \boldsymbol{X} = \bigcup \{A \cap B \mid B \in \boldsymbol{X}\}$

    (ii) $A \cup \bigcap \boldsymbol{X} = \bigcap \{A \cup B \mid B \in \boldsymbol{X}\}$

    (iii) $A \setminus \bigcup \boldsymbol{X} = \bigcap \{A \setminus B \mid B \in \boldsymbol{X}\}$

    (iv) $A \setminus \bigcap \boldsymbol{X} = \bigcup \{A \setminus B \mid B \in \boldsymbol{X}\}$.

Proofs:

(i) $x \in A \cap \bigcup \boldsymbol{X} \Leftrightarrow x \in A$ and $x \in \bigcup \boldsymbol{X} \Leftrightarrow x \in A$ and there is a $B \in \boldsymbol{X}$ with $x \in B \Leftrightarrow x \in A \cap B$ for some $B \in \boldsymbol{X} \Leftrightarrow x \in \bigcup \{A \cap B \mid B \in \boldsymbol{X}\}$. □

(ii) $x \in A \cup \bigcap \boldsymbol{X} \Leftrightarrow x \in A$ or $x \in \bigcap \boldsymbol{X} \Leftrightarrow x \in A$ or $x \in B$ for every $B \in \boldsymbol{X} \Leftrightarrow x \in A \cup B$ for every $B \in \boldsymbol{X} \Leftrightarrow x \in \bigcap \{A \cup B \mid B \in \boldsymbol{X}\}$. □

(iii) $x \in A \setminus \bigcup \boldsymbol{X} \Leftrightarrow x \in A$ and $x \notin \bigcup \boldsymbol{X} \Leftrightarrow x \in A$ and $x \notin B$ for every $B \in \boldsymbol{X} \Leftrightarrow x \in A \setminus B$ for every $B \in \boldsymbol{X} \Leftrightarrow x \in \bigcap \{A \setminus B \mid B \in \boldsymbol{X}\}$. □

(iv) $x \in A \setminus \bigcap \boldsymbol{X} \Leftrightarrow x \in A$ and $x \notin \bigcap \boldsymbol{X} \Leftrightarrow x \in A$ and $x \notin B$ for some $B \in \boldsymbol{X} \Leftrightarrow x \in A \setminus B$ for some $B \in \boldsymbol{X} \Leftrightarrow x \in \bigcup \{A \setminus B \mid B \in \boldsymbol{X}\}$. □

Note: The rules in (i) and (ii) are known as the **generalized distributive laws** and the rules in (iii) and (iv) are known as the **generalized De Morgan's laws.**

Problem Set 3

LEVEL 1

1. List the elements of $\{\Gamma, \mathrm{T}, \Delta\} \times \{\gamma, \delta\}$.

Solution: $(\Gamma,\gamma), (\Gamma,\delta), (\mathrm{T},\gamma), (\mathrm{T},\delta), (\Delta,\gamma), (\Delta,\delta)$

2. For each set A below, evaluate (i) A^2; (ii) A^3; (iii) $\mathcal{P}(A)$.

1. $A = \emptyset$ 2. $A = \{\emptyset\}$ 3. $A = \mathcal{P}(\{\emptyset\})$

Solutions:

(i) $\emptyset^2 = \emptyset \times \emptyset = \emptyset$.

$\{\emptyset\}^2 = \{\emptyset\} \times \{\emptyset\} = \{(\emptyset, \emptyset)\}$.

Since $\mathcal{P}(\{\emptyset\}) = \{\emptyset, \{\emptyset\}\}$, we have

$$\mathcal{P}(\{\emptyset\})^2 = \mathcal{P}(\{\emptyset\}) \times \mathcal{P}(\{\emptyset\}) = \{(\emptyset,\emptyset), (\emptyset,\{\emptyset\}), (\{\emptyset\},\emptyset), (\{\emptyset\},\{\emptyset\})\}.$$

(ii) $\emptyset^3 = \emptyset \times \emptyset \times \emptyset = \emptyset$.

$\{\emptyset\}^3 = \{\emptyset\} \times \{\emptyset\} \times \{\emptyset\} = \{(\emptyset,\emptyset,\emptyset)\}$.

Since $\mathcal{P}(\{\emptyset\}) = \{\emptyset, \{\emptyset\}\}$, we have $\mathcal{P}(\{\emptyset\})^3 = \mathcal{P}(\{\emptyset\}) \times \mathcal{P}(\{\emptyset\}) \times \mathcal{P}(\{\emptyset\})$

$= \{(\emptyset,\emptyset,\emptyset), (\emptyset,\emptyset,\{\emptyset\}), (\emptyset,\{\emptyset\},\emptyset), (\emptyset,\{\emptyset\},\{\emptyset\}), (\{\emptyset\},\emptyset,\emptyset), (\{\emptyset\},\emptyset,\{\emptyset\}), (\{\emptyset\},\{\emptyset\},\emptyset), (\{\emptyset\},\{\emptyset\},\{\emptyset\})\}$

(iii) $\mathcal{P}(\emptyset) = \{\emptyset\}$

$\mathcal{P}(\{\emptyset\}) = \{\emptyset, \{\emptyset\}\}$

Since $\mathcal{P}(\{\emptyset\}) = \{\emptyset, \{\emptyset\}\}$, we have $\mathcal{P}(\mathcal{P}(\{\emptyset\})) = \{\emptyset, \{\emptyset\}, \{\{\emptyset\}\}, \{\emptyset, \{\emptyset\}\}\}$.

3. Let $C = (-\infty, 2]$ and $D = (-1, 3]$. Compute each of the following:

(i) $C \cup D$

(ii) $C \cap D$

(iii) $C \setminus D$

(iv) $D \setminus C$

(v) $C \,\Delta\, D$

(i) $C \cup D = (-\infty, 3]$

(ii) $C \cap D = (-1, 2]$

(iii) $C \setminus D = (-\infty, -1]$

(iv) $D \setminus C = (2, 3]$

(v) $C \,\Delta\, D = (-\infty, -1] \cup (2, 3]$

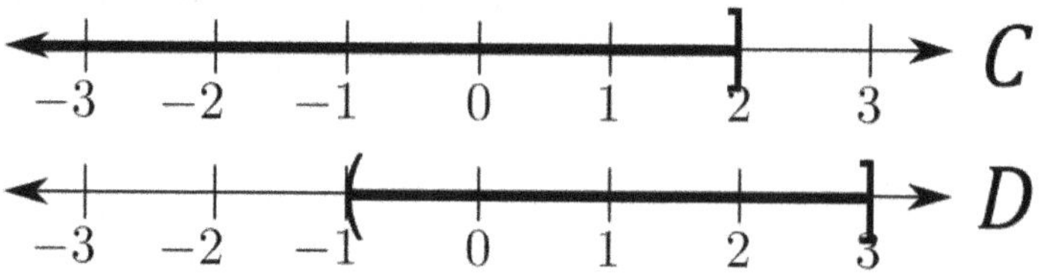

4. Find all partitions of the three-element set $\{a, b, c\}$ and the four-element set $\{a, b, c, d\}$.

Solution: The partitions of $\{a, b, c\}$ are $\{\{a\}, \{b\}, \{c\}\}$, $\{\{a\}, \{b, c\}\}$, $\{\{b\}, \{a, c\}\}$, $\{\{c\}, \{a, b\}\}$, and $\{\{a, b, c\}\}$.

The partitions of $\{a, b, c, d\}$ are $\{\{a\}, \{b\}, \{c\}, \{d\}\}$, $\{\{a\}, \{b\}, \{c, d\}\}$, $\{\{a\}, \{c\}, \{b, d\}\}$, $\{\{a\}, \{d\}, \{b, c\}\}$, $\{\{b\}, \{c\}, \{a, d\}\}$, $\{\{b\}, \{d\}, \{a, c\}\}$, $\{\{c\}, \{d\}, \{a, b\}\}$, $\{\{a, b\}, \{c, d\}\}$, $\{\{a, c\}, \{b, d\}\}$, $\{\{a, d\}, \{b, c\}\}$, $\{\{a, b, c\}, \{d\}\}$, $\{\{a, b, d\}, \{c\}\}$, $\{\{a, c, d\}, \{b\}\}$, $\{\{b, c, d\}, \{a\}\}$, and $\{\{a, b, c, d\}\}$.

5. Let $A = \{1, 2, 3, 4\}$ and let $R = \{(1,1), (1,3), (2,2), (2,4), (3,1), (3,3), (4,2), (4,4)\}$. Note that R is an equivalence relation on A. Find the equivalence classes of R.

Solution: The equivalence classes of R are $\{1, 3\}$ and $\{2, 4\}$.

LEVEL 2

6. Compute each of the following:
   (i) $\{0, 1, 2\}^3$
   (ii) $\{a, b\}^4$

Solutions:

(i) $\{0, 1, 2\}^3 = \{0, 1, 2\} \times \{0, 1, 2\} \times \{0, 1, 2\} = \{(0, 0, 0), (0, 0, 1), (0, 0, 2), (0, 1, 0), (0, 1, 1), (0, 1, 2), (0, 2, 0), (0, 2, 1), (0, 2, 2), (1, 0, 0), (1, 0, 1), (1, 0, 2), (1, 1, 0), (1, 1, 1), (1, 1, 2), (1, 2, 0), (1, 2, 1), (1, 2, 2), (2, 0, 0), (2, 0, 1), (2, 0, 2), (2, 1, 0), (2, 1, 1), (2, 1, 2), (2, 2, 0), (2, 2, 1), (2, 2, 2)\}$.

(ii) $\{a, b\}^4 = \{a, b\} \times \{a, b\} \times \{a, b\} \times \{a, b\} = \{(a, a, a, a), (a, a, a, b), (a, a, b, a), (a, a, b, b), (a, b, a, a), (a, b, a, b), (a, b, b, a), (a, b, b, b), (b, a, a, a), (b, a, a, b), (b, a, b, a), (b, a, b, b), (b, b, a, a), (b, b, a, b), (b, b, b, a), (b, b, b, b)\}$,

7. Find the domain, range, and field of each of the following relations:
   (i) $R = \{(a, b), (c, d), (e, f), (f, a)\}$
   (ii) $S = \{(2k, 2t + 1) \mid k, t \in \mathbb{Z}\}$

Solutions:

(i) $\text{dom } R = \{a, c, e, f\}$; $\text{ran } R = \{a, b, d, f\}$; $\text{field } R = \{a, b, c, d, e, f\}$

(ii) $\text{dom } S = 2\mathbb{Z} = \mathbb{E}$; $\text{ran } S = \{2t + 1 \mid t \in \mathbb{Z}\} = \mathbb{O}$; $\text{field } S = \mathbb{Z}$

8. Prove that for each $n \in \mathbb{Z}^+$, $\equiv_n$ (see part 4 of Example 3.17) is an equivalence relation on $\mathbb{Z}$.

Proof: Let $a \in \mathbb{Z}$. Then $a - a = 0 = n \cdot 0$. So, $n|a - a$. Therefore, $a \equiv_n a$, and so, $\equiv_n$ is reflexive. Let $a, b \in \mathbb{Z}$ and suppose that $a \equiv_n b$. Then $n|b - a$. So, there is $k \in \mathbb{Z}$ such that $b - a = nk$. Thus, $a - b = -(b - a) = -nk = n(-k)$. Since $k \in \mathbb{Z}$, $-k \in \mathbb{Z}$. So, $n|a - b$, and therefore, $b \equiv_n a$. So, $\equiv_n$ is symmetric. Let $a, b, c \in \mathbb{Z}$ with $a \equiv_n b$ and $b \equiv_n c$. Then $n|b - a$ and $n|c - b$. So, there are $j, k \in \mathbb{Z}$ such that $b - a = nj$ and $c - b = nk$. So, $c - a = (c - b) + (b - a) = nk + nj = n(k + j)$. Since $\mathbb{Z}$ is closed under addition, $k + j \in \mathbb{Z}$. Therefore, $n|c - a$. So, $a \equiv_n c$. Thus, $\equiv_n$ is transitive. Since $\equiv_n$ is reflexive, symmetric, and transitive, $\equiv_n$ is an equivalence relation on $\mathbb{Z}$. □

LEVEL 3

9. Let A, B, C, and D be sets with $A \subseteq B$ and $C \subseteq D$. Prove that $A \times C \subseteq B \times D$.

Proof: Let A, B, C, and D be sets with $A \subseteq B$ and $C \subseteq D$, and let $(x, y) \in A \times C$. Then $x \in A$ and $y \in C$. Since $x \in A$ and $A \subseteq B$, $x \in B$. Since $y \in C$ and $C \subseteq D$, $y \in D$. Therefore, $(x, y) \in B \times D$. Since $(x, y) \in A \times C$ was arbitrary, $A \times C \subseteq B \times D$. □

10. Prove that there do not exist sets A and B such that the relation $<$ on $\mathbb{R}$ is equal to $A \times B$.

Proof: Suppose toward contradiction that $< = A \times B$ for some sets A and B. Since $1 < 2$ and $2 < 3$, we have $(1, 2), (2, 3) \in A \times B$. Since $(2, 3) \in A \times B$, $2 \in A$. Since $(1, 2) \in A \times B$, $2 \in B$. Therefore, $(2, 2) \in A \times B$. Since $< = A \times B$, $2 < 2$, contradicting that $<$ is antireflexive. Therefore, there do not exist sets A and B such that $<$ is equal to $A \times B$. □

11. Let $\boldsymbol{X}$ be a set of equivalence relations on a nonempty set A. Prove that $\bigcap \boldsymbol{X}$ is an equivalence relation on A.

Proof: Let $\boldsymbol{X}$ be a set of equivalence relations on a nonempty set A. Let $x \in A$ and let $R \in \boldsymbol{X}$. Since R is reflexive, $(x, x) \in R$. Since $R \in \boldsymbol{X}$ was arbitrary, $\forall R \in \boldsymbol{X}((x, x) \in R)$. So, $(x, x) \in \bigcap \boldsymbol{X}$. Since $x \in A$ was arbitrary, $\bigcap \boldsymbol{X}$ is reflexive.

Let $(x, y) \in \bigcap \boldsymbol{X}$ and let $R \in \boldsymbol{X}$. Then $(x, y) \in R$. Since R is an equivalence relation, R is symmetric. Therefore, $(y, x) \in R$. Since $R \in \boldsymbol{X}$ was arbitrary, $\forall R \in \boldsymbol{X}((y, x) \in R)$. So, $(y, x) \in \bigcap \boldsymbol{X}$. Since $(x, y) \in \bigcap \boldsymbol{X}$ was arbitrary, $\bigcap \boldsymbol{X}$ is symmetric.

Let $(x, y), (y, z) \in \bigcap \boldsymbol{X}$ and let $R \in \boldsymbol{X}$. Then $(x, y), (y, z) \in R$. Since R is an equivalence relation, R is transitive. Therefore, $(x, z) \in R$. Since $R \in \boldsymbol{X}$ was arbitrary, $\forall R \in \boldsymbol{X}((x, z) \in R)$. So, $(x, z) \in \bigcap \boldsymbol{X}$. Since $(x, y), (y, z) \in \bigcap \boldsymbol{X}$ was arbitrary, $\bigcap \boldsymbol{X}$ is transitive.

Since $\bigcap \boldsymbol{X}$ is reflexive, symmetric, and transitive, $\bigcap \boldsymbol{X}$ is an equivalence relation. □

LEVEL 4

12. Let $R = \{(x, y) \in \mathbb{R} \times \mathbb{R} \mid x - y \in \mathbb{Z}\}$. Prove that R is an equivalence relation on $\mathbb{R}$ and describe the equivalence classes of R.

Proof: If $x \in \mathbb{R}$, then $x - x = 0 \in \mathbb{Z}$. So, xRx, and therefore, R is reflexive. If xRy, then $x - y \in \mathbb{Z}$. It follows that $y - x = -(x - y) \in \mathbb{Z}$. So, yRx, and therefore, R is symmetric. If xRy and yRz, then $x - y \in \mathbb{Z}$ and $y - z \in \mathbb{Z}$. It follows that $x - z = (x - y) + (y - z) \in \mathbb{Z}$ (because the sum of two integers is an integer). So, xRy, and therefore, R is transitive. Since R is reflexive, symmetric, and transitive, R is an equivalence relation.

For each $r \in \mathbb{R}$ with $0 \leq r < 1$, the set $X_r = \{r + n \mid n \in \mathbb{Z}\}$ is an equivalence class of R. To see this, first note that if $n, m \in \mathbb{Z}$, then $(r + n) - (r + m) = n - m \in \mathbb{Z}$. So, any two elements of X_r are equivalent. Also, if x is equivalent to $r + n$, then there is an integer m so that $(r + n) - x = m$. It follows that $x = r + (n - m)$, and so, $x \in X_r$.

Now, if $x \in \mathbb{R}$, then let n be an integer with $n \leq x < n + 1$. Then $0 \leq x - n < 1$. Let $r = x - n$. We have $x - r = x - (x - n) = n$, and so, $x = r + n$. Therefore, $x \in X_r$.

Finally, if $0 \leq r < 1$ and $0 \leq s < 1$ with $r \leq s$ and rRs, then we have $s - r \geq 0$ and $s - r < 1$. Therefore, $s - r = 0$, and so, $s = r$. □

13. Let A, B, C, and D be sets. Determine if each of the following statements is true or false. If true, provide a proof. If false, provide a counterexample.

    (i) $(A \times B) \cap (C \times D) = (A \cap C) \times (B \cap D)$

    (ii) $(A \times B) \cup (C \times D) = (A \cup C) \times (B \cup D)$

Solutions:

(i) This is **true**.

Proof: $(x, y) \in (A \times B) \cap (C \times D)$ if and only if $(x, y) \in A \times B$ and $(x, y) \in C \times D$ if and only if $x \in A$, $y \in B$, $x \in C$, and $y \in D$ if and only if $x \in A \cap C$ and $y \in B \cap D$ if and only if $(x, y) \in (A \cap C) \times (B \cap D)$. Therefore, $(A \times B) \cap (C \times D) = (A \cap C) \times (B \cap D)$. □

(ii) This is **false**. If $A = \{0\}, B = \{1\}, C = \{2\}, D = \{3\}$, then $A \times B = \{(0,1)\}$, $C \times D = \{(2,3)\}$, and so, $(A \times B) \cup (C \times D) = \{(0,1), (2,3)\}$. Also, $A \cup C = \{0,2\}$, $B \cup D = \{1,3\}$, and so, $(A \cup C) \times (B \cup D) = \{(0,1), (0,3), (2,1), (2,3)\}$. Since $(2,1) \in (A \cup C) \times (B \cup D)$, but $(2,1) \notin (A \times B) \cup (C \times D)$, we see that $(A \times B) \cup (C \times D) \neq (A \cup C) \times (B \cup D)$.

14. Let R be a relation on a set A . Determine if each of the following statements is true or false. If true, provide a proof. If false, provide a counterexample.

    (i) If R is symmetric and transitive on A, then R is reflexive on A.

    (ii) If R is antisymmetric on A, then R is not symmetric on A.

Solutions:

(i) This is **false**. Let $A = \{0,1\}$ and $R = \{(0,0)\}$. Then R is symmetric and transitive, but not reflexive (because $(1,1) \notin R$).

(ii) This is **false**. $\emptyset$ is both symmetric and antisymmetric on any set A.

LEVEL 5

15. For $a, b \in \mathbb{N}$, we will say that a divides b, written $a|b$, if there is a natural number k such that $b = ak$. Notice that $|$ is a binary relation on $\mathbb{N}$. Prove that $(\mathbb{N}, |\,)$ is a partially ordered set, but it is not a linearly ordered set.

Proof: If $a \in \mathbb{N}$ then $a = 1a$, so that $a|a$. Therefore, $|$ is reflexive. If $a|b$ and $b|a$, then there are natural numbers j and k such that $b = ja$ and $a = kb$. If $a = 0$, then $b = j \cdot 0 = 0$, and so, $a = b$. Suppose $a \neq 0$. We have $a = k(ja) = (kj)a$. Thus, $(kj - 1)a = (kj)a - 1a = 0$. So, $kj - 1 = 0$, and therefore, $kj = 1$. So, $k = j = 1$. Thus, $b = ja = 1a = a$. Therefore, $|$ is antisymmetric. If $a|b$ and $b|c$, then there are natural numbers j and k such that $b = ja$ and $c = kb$. Then $c = kb = k(ja) = (kj)a$. Since the product of two natural numbers is a natural number, $kj \in \mathbb{N}$. So, $a|c$. Therefore, $|$ is transitive. Since $|$ is reflexive, antisymmetric, and transitive on $\mathbb{N}$, $(\mathbb{N}, |\,)$ is a partially ordered set. Since 2 and 3 do not divide each other, $(\mathbb{N}, |\,)$ is **not** linearly ordered. □

16. Let $\boldsymbol{P}$ be a partition of a set S. Prove that there is an equivalence relation $\sim$ on S for which the elements of $\boldsymbol{P}$ are the equivalence classes of $\sim$. Conversely, if $\sim$ is an equivalence relation on a set S, prove that the equivalence classes of $\sim$ form a partition of S.

Proof: Let $\boldsymbol{P}$ be a partition of S, and define the relation $\sim$ by $x \sim y$ if and only if there is $X \in \boldsymbol{P}$ with $x, y \in X$.

Let $x \in S$. Since $\boldsymbol{P}$ is a partition of S, $S = \bigcup \boldsymbol{P}$. So, there is $X \in \boldsymbol{P}$ with $x \in X$. It follows that $x \sim x$. Therefore, $\sim$ is reflexive.

If $x \sim y$, then there is $X \in \boldsymbol{P}$ with $x, y \in X$. So, $y, x \in X$ (obviously!). Thus, $y \sim x$, and therefore, $\sim$ is symmetric.

If $x \sim y$ and $y \sim z$, then there are $X, Y \in \boldsymbol{P}$ with $x, y \in X$ and $y, z \in Y$. Since $y \in X$ and $y \in Y$, we have $y \in X \cap Y$. Since $\boldsymbol{P}$ is a partition and $X \cap Y \neq \emptyset$, we must have $X = Y$. So, $z \in X$. Thus, $x, z \in X$, and therefore, $x \sim z$. So, $\sim$ is transitive.

Since $\sim$ is reflexive, symmetric, and transitive on S, $\sim$ is an equivalence relation on S.

We still need to show that $\boldsymbol{P} = \{[x] \mid x \in S\}$. Let $X \in \boldsymbol{P}$ and let $x \in X$. We show that $X = [x]$. Let $y \in X$. Since $x, y \in X$, $x \sim y$. So $y \in [x]$. Thus, $X \subseteq [x]$. Now, let $y \in [x]$. Then $x \sim y$. So, there is $Y \in \boldsymbol{P}$ such that $x, y \in Y$. Since $x \in X$ and $x \in Y$, $x \in X \cap Y$. Since $\boldsymbol{P}$ is a partition and $X \cap Y \neq \emptyset$, we must have $X = Y$. So, $y \in X$. Thus, $[x] \subseteq X$. Since $X \subseteq [x]$ and $[x] \subseteq X$, we have $X = [x]$. Since $X \in \boldsymbol{P}$ was arbitrary, we have shown $\boldsymbol{P} \subseteq \{[x] \mid x \in S\}$.

Now, let $X \in \{[x] \mid x \in S\}$. Then there is $x \in S$ such that $X = [x]$. Since $\boldsymbol{P}$ is a partition of S, $S = \bigcup \boldsymbol{P}$. So, there is $Y \in \boldsymbol{P}$ with $x \in Y$. We will show that $X = Y$. Let $y \in X$. Then $x \sim y$. So, there is $Z \in \boldsymbol{P}$ with $x, y \in Z$. Since $x \in Y$ and $x \in Z$, $x \in Y \cap Z$. Since $\boldsymbol{P}$ is a partition and $Y \cap Z \neq \emptyset$, we must have $Y = Z$. So, $y \in Y$. Since $y \in X$ was arbitrary, $X \subseteq Y$. Now, let $y \in Y$. Then $x \sim y$. So, $y \in [x] = X$. Since $y \in Y$ was arbitrary, $Y \subseteq X$. Since $X \subseteq Y$ and $Y \subseteq X$, we have $X = Y$. Therefore, $X \in \boldsymbol{P}$. Since $X \in \{[x] \mid x \in S\}$ was arbitrary, we have $\{[x] \mid x \in S\} \subseteq \boldsymbol{P}$.

Since $\boldsymbol{P} \subseteq \{[x] \mid x \in S\}$ and $\{[x] \mid x \in S\} \subseteq \boldsymbol{P}$, we have $\boldsymbol{P} = \{[x] \mid x \in S\}$, as desired.

Now, let $\sim$ be an equivalence relation on S. We first show that $\bigcup\{[x] \mid x \in S\} = S$.

Let $y \in \bigcup\{[x] \mid x \in S\}$. Then there is $x \in S$ with $y \in [x]$. By definition of $[x]$, $y \in S$. Therefore, $\bigcup\{[x] \mid x \in S\} \subseteq S$. Now, let $y \in S$. Since $\sim$ is an equivalence relation, $y \sim y$. So, $y \in [y]$. Thus, $y \in \bigcup\{[x] \mid x \in S\}$. So, we have $S \subseteq \bigcup\{[x] \mid x \in S\}$. Since $\bigcup\{[x] \mid x \in S\} \subseteq S$ and $S \subseteq \bigcup\{[x] \mid x \in S\}$, $\bigcup\{[x] \mid x \in S\} = S$.

We next show that if $x, y \in S$, then $[x] \cap [y] = \emptyset$ or $[x] = [y]$.

Suppose $[x] \cap [y] \neq \emptyset$ and let $z \in [x] \cap [y]$. Then $x \sim z$ and $y \sim z$. Since $\sim$ is symmetric, $z \sim y$. Since $\sim$ is transitive, $x \sim y$. Let $w \in [x]$. Then $x \sim w$. By symmetry, $y \sim x$. By transitivity, $y \sim w$. So, $w \in [y]$. Since $w \in [x]$ was arbitrary, $[x] \subseteq [y]$. By a symmetric argument, $[y] \subseteq [x]$.

Since $[x] \subseteq [y]$ and $[y] \subseteq [x]$, we have $[x] = [y]$.

Since $\bigcup\{[x] \mid x \in S\} = S$ and every pair of equivalence classes are either disjoint or equal, the set of equivalence classes partitions S. □

Problem Set 4

LEVEL 1

1. Determine if each of the following relations are functions. For each such function, determine if it is injective. State the domain and range of each function.
   (i) $R = \{(a,b),(b,b),(c,d),(e,a)\}$
   (ii) $S = \{(a,a),(a,b),(b,a)\}$
   (iii) $T = \{(a,b) \mid a,b \in \mathbb{R} \wedge b < 0 \wedge a^2 + b^2 = 9\}$

Solutions:

(i) R is a function. It is **not** injective. $\operatorname{dom} R = \{a,b,c,e\}$ and $\operatorname{ran} R = \{a,b,d\}$.

(ii) S **is not** a function.

(iii) T is a function. It is **not** injective. $\operatorname{dom} T = (-3,3)$ and $\operatorname{ran} T = (-3,0)$.

2. Define $f: \mathbb{Z} \to \mathbb{Z}$ by $f(n) = n^2$. Let $A = \{0,1,2,3,4\}$, $B = \mathbb{N}$, and $C = \{-2n \mid n \in \mathbb{N}\}$. Evaluate each of the following:
   (i) $f[A]$
   (ii) $f^{-1}[A]$
   (iii) $f^{-1}[B]$
   (iv) $f^{-1}[B \cup C]$

Solutions:

(i) $f[A] = \{0,1,4,9,16\}$.

(ii) $f^{-1}[A] = \{-2,-1,0,1,2\}$.

(iii) $f^{-1}[B] = \mathbb{Z}$.

(iv) $f^{-1}[B \cup C] = \mathbb{Z}$.

3. Let A, B, and C be sets. Prove the following:
   (i) $\preccurlyeq$ is transitive.
   (ii) $\prec$ is transitive.
   (iii) If $A \preccurlyeq B$ and $B \prec C$, then $A \prec C$.
   (iv) If $A \prec B$ and $B \preccurlyeq C$, then $A \prec C$.

Proofs:

(i) Suppose that $A \preccurlyeq B$ and $B \preccurlyeq C$. Then there are functions $f: A \hookrightarrow B$ and $g: B \hookrightarrow C$. By Theorem 4.6, $g \circ f: A \hookrightarrow C$. So, $A \preccurlyeq C$. Therefore, $\preccurlyeq$ is transitive. □

(ii) Suppose that $A \prec B$ and $B \prec C$. Then $A \preccurlyeq B$ and $B \preccurlyeq C$. By (i), $A \preccurlyeq C$. Assume toward contradiction that $A \sim C$. Since $\sim$ is symmetric, $C \sim A$. In particular, $C \preccurlyeq A$. Since $C \preccurlyeq A$ and $A \preccurlyeq B$, by (i), $C \preccurlyeq B$. Since $B \preccurlyeq C$ and $C \preccurlyeq B$, by the Cantor-Schroeder-Bernstein Theorem, $B \sim C$, contradicting $B \prec C$. It follows that $A \nsim C$, and thus, $A \prec C$. □

(iii) Suppose that $A \preccurlyeq B$ and $B \prec C$. Then $B \preccurlyeq C$. By (i), $A \preccurlyeq C$. Assume toward contradiction that $A \sim C$. The rest of the argument is the same as (ii). □

(iv) Suppose that $A \prec B$ and $B \preccurlyeq C$. Then $A \preccurlyeq B$. By (i), $A \preccurlyeq C$. Assume toward contradiction that $A \sim C$. Since $\sim$ is symmetric, $C \sim A$. In particular, $C \preccurlyeq A$. Since $B \preccurlyeq C$ and $C \preccurlyeq A$, by (i), $B \preccurlyeq A$. Since $A \preccurlyeq B$ and $B \preccurlyeq A$, by the Cantor-Schroeder-Bernstein Theorem, $A \sim B$, contradicting $A \prec B$. It follows that $A \nsim C$, and thus, $A \prec C$. □

LEVEL 2

4. Find sets A and B and a function f such that $f[A \cap B] \neq f[A] \cap f[B]$.

Solution: Define $f: \{a, b\} \to \{0\}$ by $\{(a, 0), (b, 0)\}$. Let $A = \{a\}$ and $B = \{b\}$. Then $A \cap B = \emptyset$. Therefore, $f[A \cap B] = \emptyset$ and $f[A] \cap f[B] = \{0\} \cap \{0\} = \{0\}$.

5. Let $f: A \to B$ and let $V \subseteq B$. Prove that $f\left[f^{-1}[V]\right] \subseteq V$.

Proof: Let $y \in f\left[f^{-1}[V]\right]$. Then there is $x \in f^{-1}[V]$ with $y = f(x)$. Since $x \in f^{-1}[V]$, we have $y = f(x) \in V$. Since $y \in f\left[f^{-1}[V]\right]$ was arbitrary, $f\left[f^{-1}[V]\right] \subseteq V$. □

6. Define $\mathcal{P}_k(\mathbb{N})$ for each $k \in \mathbb{N}$ by $\mathcal{P}_0(\mathbb{N}) = \mathbb{N}$ and $\mathcal{P}_{k+1}(\mathbb{N}) = \mathcal{P}(\mathcal{P}_k(\mathbb{N}))$ for $k > 0$. Find a set B such that for all $k \in \mathbb{N}$, $\mathcal{P}_k(\mathbb{N}) \prec B$.

Solution: Let $B = \bigcup\{\mathcal{P}_n(\mathbb{N}) \mid n \in \mathbb{N}\}$. Let $k \in \mathbb{N}$. Since $\mathcal{P}_k(\mathbb{N}) \subseteq B$, by Note 1 following Example 4.16, $\mathcal{P}_k(\mathbb{N}) \preccurlyeq B$. Since k was arbitrary, we have $\mathcal{P}_k(\mathbb{N}) \preccurlyeq B$ for all $k \in \mathbb{N}$. Again, let $k \in \mathbb{N}$. We have $\mathcal{P}_k(\mathbb{N}) \prec \mathcal{P}_{k+1}(\mathbb{N})$ and $\mathcal{P}_{k+1}(\mathbb{N}) \preccurlyeq B$. By Problem 3 (part (iv)), $\mathcal{P}_k(\mathbb{N}) \prec B$. Since $k \in \mathbb{N}$ was arbitrary, we have shown that for all $k \in \mathbb{N}$, $\mathcal{P}_k(\mathbb{N}) \prec B$.

7. Prove that if $A \sim B$ and $C \sim D$, then $A \times C \sim B \times D$.

Proof: Suppose that $A \sim B$ and $C \sim D$. Then there exist bijections $h: A \to B$ and $k: C \to D$. Define $f: A \times C \to B \times D$ by $f(a, c) = (h(a), k(c))$.

Suppose $(a, c), (a', c') \in A \times C$ with $f((a, c)) = f((a', c'))$. Then $(h(a), k(c)) = (h(a'), k(c'))$. So, $h(a) = h(a')$ and $k(c) = k(c')$. Since h is an injection, $a = a'$. Since k is an injection, $c = c'$. Since $a = a'$ and $c = c'$, $(a, c) = (a', c')$. Since $(a, c), (a', c') \in A \times C$ were arbitrary, f is an injection.

Now, let $(b, d) \in B \times D$. Since h and k are bijections, h^{-1} and k^{-1} exist. Let $a = h^{-1}(b)$, $c = k^{-1}(d)$. Then $f(a, c) = (h(a), k(c)) = \left(h(h^{-1}(b)), k(k^{-1}(d))\right) = (b, d)$. Since $(b, d) \in B \times D$ was arbitrary, f is a surjection.

Since f is both an injection and a surjection, $A \times C \sim B \times D$. □

LEVEL 3

8. For $f, g \in {}^{\mathbb{R}}\mathbb{R}$, define $f \preccurlyeq g$ if and only if for all $x \in \mathbb{R}$, $f(x) \le g(x)$. Is $({}^{\mathbb{R}}\mathbb{R}, \preccurlyeq)$ a poset? Is it a linearly ordered set? What if we replace $\preccurlyeq$ by $\preccurlyeq^*$, where $f \preccurlyeq^* g$ if and only if there is an $x \in \mathbb{R}$ such that $f(x) \le g(x)$?

Solution: If $f \in {}^{\mathbb{R}}\mathbb{R}$, then for all $x \in \mathbb{R}$, $f(x) = f(x)$. So, $f \preccurlyeq f$, and therefore, $\preccurlyeq$ is reflexive.

Let $f, g \in {}^{\mathbb{R}}\mathbb{R}$ with $f \preccurlyeq g$ and $g \preccurlyeq f$. Then for all $x \in \mathbb{R}$, $f(x) \le g(x)$ and $g(x) \le f(x)$. So, $f = g$, and therefore, $\preccurlyeq$ is antisymmetric.

Let $f, g, h \in {}^{\mathbb{R}}\mathbb{R}$ with $f \preccurlyeq g$ and $g \preccurlyeq h$. Then for all $x \in \mathbb{R}$, $f(x) \le g(x)$ and $g(x) \le h(x)$. So, by the transitivity of $\le$, for all $x \in \mathbb{R}$, $f(x) \le h(x)$. Thus, $f \preccurlyeq h$, and therefore, $\preccurlyeq$ is transitive.

Since $\preccurlyeq$ is reflexive, antisymmetric, and transitive, $({}^{\mathbb{R}}\mathbb{R}, \preccurlyeq)$ is a poset.

Let $f(x) = x$ and $g(x) = x^2$. Then $f(2) = 2$ and $g(2) = 4$. So, $f(2) < g(2)$. Therefore, $g \not\preccurlyeq f$. We also have $f\left(\frac{1}{2}\right) = \frac{1}{2}$ and $g\left(\frac{1}{2}\right) = \frac{1}{4}$. So, $g\left(\frac{1}{2}\right) < f\left(\frac{1}{2}\right)$. Therefore, $f \not\preccurlyeq g$. So, f and g are incomparible with respect to $\preccurlyeq$. Therefore, $({}^{\mathbb{R}}\mathbb{R}, \preccurlyeq)$ is **not** a linearly ordered set.

The same example from the last paragraph gives us $f \preccurlyeq^* g$ and $g \preccurlyeq^* f$. But $f \ne g$. So, $\preccurlyeq^*$ is **not** antisymmetric, and therefore, $({}^{\mathbb{R}}\mathbb{R}, \preccurlyeq^*)$ is **not** a poset.

9. Prove that the function $f: \mathbb{N} \to \mathbb{Z}$ defined by $f(n) = \begin{cases} \frac{n}{2} & \text{if } n \text{ is even} \\ -\frac{n+1}{2} & \text{if } n \text{ is odd} \end{cases}$ is a bijection.

Proof: First note that if n is even, then there is $k \in \mathbb{Z}$ with $n = 2k$, and so, $\frac{n}{2} = \frac{2k}{2} = k \in \mathbb{Z}$, and if n is odd, there is $k \in \mathbb{Z}$ with $n = 2k + 1$, and so, $-\frac{n+1}{2} = -\frac{(2k+1)+1}{2} = -\frac{2k+2}{2} = -\frac{2(k+1)}{2} = -(k+1) \in \mathbb{Z}$. So, f does take each natural number to an integer.

Now, suppose that $n, m \in \mathbb{N}$ with $f(n) = f(m)$. If n and m are both even, we have $\frac{n}{2} = \frac{m}{2}$, and so, $2 \cdot \frac{n}{2} = 2 \cdot \frac{m}{2}$. Thus, $n = m$. If n and m are both odd, we have $-\frac{n+1}{2} = -\frac{m+1}{2}$, and so, $\frac{n+1}{2} = \frac{m+1}{2}$. Thus, $2 \cdot \frac{n+1}{2} = 2 \cdot \frac{m+1}{2}$. So, $n + 1 = m + 1$, and therefore, $n = m$. If n is even and m is odd, then we have $\frac{n}{2} = -\frac{m+1}{2}$. So, $2 \cdot \frac{n}{2} = 2\left(-\frac{m+1}{2}\right)$. Therefore, $n = -(m+1)$. Since $m \in \mathbb{N}$, $m \ge 0$. So, $m + 1 \ge 1$. Therefore, $n = -(m+1) \le -1$, contradicting $n \in \mathbb{N}$. So, it is impossible for n to be even, m to be odd, and $f(n) = f(m)$. Similarly, we cannot have n odd and m even. So, f is an injection.

Now, let $k \in \mathbb{Z}$. If $k \ge 0$, then $2k \in \mathbb{N}$ and $f(2k) = \frac{2k}{2} = k$. If $k < 0$, then $-2k > 0$, and so, we have $-2k - 1 \in \mathbb{N}$. Then $f(-2k - 1) = -\frac{(-2k-1)+1}{2} = -\frac{-2k}{2} = k$. So, f is a surjection.

Since f is both an injection and a surjection, f is a bijection. □

10. Define a partition $\boldsymbol{P}$ of $\mathbb{N}$ such that $\boldsymbol{P} \sim \mathbb{N}$ and for each $X \in \boldsymbol{P}$, $X \sim \mathbb{N}$.

Proof: For each $n \in \mathbb{N}$, let P_n be the set of natural numbers ending with exactly n zeros and let $\boldsymbol{P} = \{P_n \mid n \in \mathbb{N}\}$. For example, $5231 \in P_0$, $0 \in P_1$, and $26{,}200 \in P_2$. Let's define $\widetilde{m,n}$ to be the natural number consisting of m 1's followed by n 0's. For example, $\widetilde{3,0} = 111$ and $\widetilde{2,5} = 1{,}100{,}000$. For each $n \in \mathbb{N}$, $\{\widetilde{m,n} \mid m \in \mathbb{N}\} \subseteq P_n$ showing that each P_n is equinumerous to $\mathbb{N}$. Also, if $k \in P_n \cap P_m$, then k ends with exactly n zeros and exactly m zeros, and so, $n = m$. Therefore, $\boldsymbol{P}$ is pairwise disjoint. This also shows that the function $f: \mathbb{N} \to \boldsymbol{P}$ defined by $f(n) = P_n$ is a bijection. So, $\boldsymbol{P} \sim \mathbb{N}$. Finally, if $k \in \mathbb{N}$, then there is $n \in \mathbb{N}$ such that k ends with exactly n zeros. So, $\bigcup \boldsymbol{P} = \mathbb{N}$. □

11. Prove that a countable union of countable sets is countable.

Proof: For each $n \in \mathbb{N}$, let A_n be a countable set. By replacing each A_n by $A_n \times \{n\}$, we can assume that $\{A_n \mid n \in \mathbb{N}\}$ is a pairwise disjoint collection of sets ($A_n \sim A_n \times \{n\}$ via the bijection f sending x to (x, n)). By Problem 10, there is a partition $\boldsymbol{P}$ of $\mathbb{N}$ such that $\boldsymbol{P} \sim \mathbb{N}$ and for each $X \in \boldsymbol{P}$, $X \sim \mathbb{N}$. Let's say $\boldsymbol{P} = \{P_n \mid n \in \mathbb{N}\}$. Since each A_n is countable, for each $n \in \mathbb{N}$ there are injections $f_n: A_n \to P_n$. Define $f: \bigcup\{A_n | n \in \mathbb{N}\} \to \mathbb{N}$ by $f(x) = f_n(x)$ if $x \in A_n$.

Since $\{A_n \mid n \in \mathbb{N}\}$ is pairwise disjoint, f is well-defined.

Suppose that $x, y \in \bigcup\{A_n | n \in \mathbb{N}\}$ with $f(x) = f(y)$. There exist $n, m \in \mathbb{N}$ such that $x \in A_n$ and $y \in A_m$. So, $f(x) = f_n(x) \in P_n$ and $f(y) = f_m(y) \in P_m$. Since $f(x) = f(y)$, we have $f_n(x) = f_m(y)$. Since for $n \neq m$, $P_n \cap P_m = \emptyset$, we must have $n = m$. So, we have $f_n(x) = f_n(y)$. Since f_n is injective, $x = y$. Since $x, y \in \bigcup\{A_n | n \in \mathbb{N}\}$ were arbitrary, f is an injective function. Therefore, $\bigcup\{A_n | n \in \mathbb{N}\}$ is countable. □

12. Let A and B be sets such that $A \sim B$. Prove that $\mathcal{P}(A) \sim \mathcal{P}(B)$.

Proof: Suppose that $A \sim B$. Then there exists a bijection $h: A \to B$. Define $F: \mathcal{P}(A) \to \mathcal{P}(B)$ by $F(X) = \{h(a) \mid a \in X\}$ for each $X \in \mathcal{P}(A)$.

Suppose $X, Y \in \mathcal{P}(A)$ with $F(X) = F(Y)$. Let $a \in X$. Then $h(a) \in F(X)$. Since $F(X) = F(Y)$, $h(a) \in F(Y)$. So, there is $b \in Y$ such that $h(a) = h(b)$. Since h is injective, $a = b$. So, $a \in Y$. Since $a \in X$ was arbitrary, $X \subseteq Y$. By a symmetrical argument, $Y \subseteq X$. Therefore, $X = Y$. Since $X, Y \in \mathcal{P}(A)$ were arbitrary, F is injective.

Let $Y \in \mathcal{P}(B)$, and let $X = \{a \in A \mid h(a) \in Y\}$. Then $b \in F(X)$ if and only if $b = h(a)$ for some $a \in X$ if and only if $b \in Y$ (because h is surjective). So, $F(X) = Y$. Since $Y \in \mathcal{P}(B)$ was arbitrary, F is surjective.

Since F is injective and surjective, $\mathcal{P}(A) \sim \mathcal{P}(B)$. □

LEVEL 4

13. Prove the following:

(i) $\mathbb{N} \times \mathbb{N} \sim \mathbb{N}$.

(ii) $\mathbb{Q} \sim \mathbb{N}$.

(iii) Any two intervals of real numbers are equinumerous (including $\mathbb{R}$ itself).

(iv) ${}^{\mathbb{N}}\mathbb{N} \sim \mathcal{P}(\mathbb{N})$.

Proofs:

(i) $\mathbb{N} \times \mathbb{N} = \bigcup\{\mathbb{N} \times \{n\} \mid n \in \mathbb{N}\}$. This is a countable union of countable sets. By Problem 11, $\mathbb{N} \times \mathbb{N}$ is countable. □

(ii) $\mathbb{Q}^+ = \left\{\frac{a}{b} \,\middle|\, a \in \mathbb{N} \wedge b \in \mathbb{N}^+\right\} = \bigcup\left\{\left\{\frac{a}{b} \,\middle|\, a \in \mathbb{N}\right\} \,\middle|\, b \in \mathbb{N}^+\right\}$. This is a countable union of countable sets. By Problem 11, $\mathbb{Q}^+$ is countable. Now, $\mathbb{Q} = \mathbb{Q}^+ \cup \{0\} \cup \mathbb{Q}^-$, where $\mathbb{Q}^- = \{q \in \mathbb{Q} \mid -q \in \mathbb{Q}^+\}$. This is again a countable union of countable sets, thus countable. So, $\mathbb{Q} \sim \mathbb{N}$. □

(iii) The function $f\colon \mathbb{R} \to (0,\infty)$ defined by $f(x) = 2^x$ is a bijection. So, $\mathbb{R} \sim (0,\infty)$. The function $g\colon (0,\infty) \to (0,1)$ defined by $g(x) = \frac{1}{x^2+1}$ is a bijection. So, $(0,\infty) \sim (0,1)$. If $a, b \in \mathbb{R}$, the function $h\colon (0,1) \to (a,b)$ defined by $h(x) = (b-a)x + a$ is a bijection. So, $(0,1) \sim (a,b)$. It follows that all bounded open intervals are equinumerous with each other and $\mathbb{R}$.

We have, $[a,b] \subseteq (a-1,b+1) \sim (a,b) \subseteq [a,b) \subseteq [a,b]$ and $(a,b) \subseteq (a,b] \subseteq [a,b]$. It follows that all bounded intervals are equinumerous with each other and $\mathbb{R}$.

We also have the following.

$$(a,\infty) \subseteq [a,\infty) \subseteq \mathbb{R} \sim (a,a+1) \subseteq (a,\infty)$$

$$(-\infty,b) \subseteq (-\infty,b] \subseteq \mathbb{R} \sim (b-1,b) \subseteq (-\infty,b)$$

Therefore, all unbounded intervals are equinumerous with $\mathbb{R}$. It follows that any two intervals of real numbers are equinumerous. □

(iv) ${}^{\mathbb{N}}\mathbb{N} \subseteq \mathcal{P}(\mathbb{N} \times \mathbb{N})$ by the definition of ${}^{\mathbb{N}}\mathbb{N}$. So, ${}^{\mathbb{N}}\mathbb{N} \preccurlyeq \mathcal{P}(\mathbb{N} \times \mathbb{N})$ by Note 1 following Example 4.16. By (i) above, $\mathbb{N} \times \mathbb{N} \sim \mathbb{N}$. So, by Problem 12, $\mathcal{P}(\mathbb{N} \times \mathbb{N}) \sim \mathcal{P}(\mathbb{N})$. So, $\mathcal{P}(\mathbb{N} \times \mathbb{N}) \preccurlyeq \mathcal{P}(\mathbb{N})$. Since $\preccurlyeq$ is transitive, ${}^{\mathbb{N}}\mathbb{N} \preccurlyeq \mathcal{P}(\mathbb{N})$.

Now, $\mathcal{P}(\mathbb{N}) \sim {}^{\mathbb{N}}\{0,1\}$ (see Example 4.11 (part 5)). So, $\mathcal{P}(\mathbb{N}) \preccurlyeq {}^{\mathbb{N}}\{0,1\}$. Also, ${}^{\mathbb{N}}\{0,1\} \subseteq {}^{\mathbb{N}}\mathbb{N}$, and so, by Note 1 following Example 4.16, ${}^{\mathbb{N}}\{0,1\} \preccurlyeq {}^{\mathbb{N}}\mathbb{N}$. Since $\preccurlyeq$ is transitive, $\mathcal{P}(\mathbb{N}) \preccurlyeq {}^{\mathbb{N}}\mathbb{N}$.

By the Cantor-Schroeder-Bernstein Theorem, ${}^{\mathbb{N}}\mathbb{N} \sim \mathcal{P}(\mathbb{N})$. □

Notes: (1) In the proof of (iii), we used the fact that equinumerosity is an equivalence relation, the Cantor-Schroeder-Bernstein Theorem, and Note 1 following Example 4.16 many times without mention. For example, we have $\mathbb{R} \sim (0, \infty)$ and $(0, \infty) \sim (0, 1)$. So, by the transitivity of $\sim$, we have $\mathbb{R} \sim (0, 1)$. As another example, the sequence $(a, \infty) \subseteq [a, \infty) \subseteq \mathbb{R} \sim (a, a+1) \subseteq (a, \infty)$ together with Note 1 following Example 4.16 gives us that $(a, \infty) \preccurlyeq \mathbb{R}$ and $\mathbb{R} \preccurlyeq (a, \infty)$. By the Cantor-Schroeder-Bernstein Theorem, $(a, \infty) \sim \mathbb{R}$.

(2) Once we showed that for all $a, b \in \mathbb{R}$, $(0, 1) \sim (a, b)$, it follows from the fact that $\sim$ is an equivalence relation that any two bounded open intervals are equinumerous. Indeed, if (a, b) and (c, d) are bounded open intervals, then $(0, 1) \sim (a, b)$ and $(0, 1) \sim (c, d)$. By the symmetry of $\sim$, we have $(a, b) \sim (0, 1)$, and finally, by the transitivity of $\sim$, we have $(a, b) \sim (c, d)$.

(3) It's easy to prove that two specific intervals of real numbers are equinumerous using just the fact that any two bounded open intervals are equinumerous with each other, together with the fact that $\mathbb{R} \sim (0, 1)$. For example, to show that $[3, \infty)$ is equinumerous with $(-2, 5]$, simply consider the following sequence: $[3, \infty) \subseteq \mathbb{R} \sim (0, 1) \sim (-2, 5) \subseteq (-2, 5] \subseteq (-2, 6) \sim (3, 4) \subseteq [3, \infty)$.

14. Prove that if $A \sim B$ and $C \sim D$, then ${}^AC \sim {}^BD$.

Proof: Suppose that $A \sim B$ and $C \sim D$. Then there exist bijections $h: A \to B$ and $k: C \to D$. Define $F: {}^AC \to {}^BD$ by $F(f)(b) = k\left(f(h^{-1}(b))\right)$.

Suppose $f, g \in {}^AC$ with $F(f) = F(g)$. Let $a \in A$, and let $b = h(a)$. We have $F(f)(b) = F(g)(b)$, or equivalently, $k\left(f(h^{-1}(b))\right) = k\left(g(h^{-1}(b))\right)$. Since k is injective, $f(h^{-1}(b)) = g(h^{-1}(b))$. Since $b = h(a)$, $a = h^{-1}(b)$. So, $f(a) = g(a)$. Since $a \in A$ was arbitrary, $f = g$. Since $f, g \in {}^AC$ were arbitrary, F is injective.

Now, let $g \in {}^BD$ and let's define $f \in {}^AC$ by $f(a) = k^{-1}\left(g(h(a))\right)$. Let $b \in B$. Then we have $F(f)(b) = k\left(f(h^{-1}(b))\right) = k\left(k^{-1}\left(g\left(h(h^{-1}(b))\right)\right)\right) = g(b)$. Since $b \in B$ was arbitrary, we have $F(f) = g$. Since $g \in {}^BD$ was arbitrary, F is surjective.

Since F is injective and surjective, ${}^AC \sim {}^BD$. □

LEVEL 5

15. Let $\boldsymbol{X}$ be a nonempty set of sets and let f be a function such that $\bigcup \boldsymbol{X} \subseteq \operatorname{dom} f$. Prove each of the following:

   (i) $f[\bigcup \boldsymbol{X}] = \bigcup\{f[A] \mid A \in \boldsymbol{X}\}$
   (ii) $f[\bigcap \boldsymbol{X}] \subseteq \bigcap\{f[A] \mid A \in \boldsymbol{X}\}$
   (iii) $f^{-1}[\bigcup \boldsymbol{X}] = \bigcup\{f^{-1}[A] \mid A \in \boldsymbol{X}\}$
   (iv) $f^{-1}[\bigcap \boldsymbol{X}] = \bigcap\{f^{-1}[A] \mid A \in \boldsymbol{X}\}$

Proofs:

(i) Let $y \in f[\bigcup \boldsymbol{X}]$. Then there is $x \in \bigcup \boldsymbol{X}$ such that $f(x) = y$. Since $x \in \bigcup \boldsymbol{X}$, there is $B \in \boldsymbol{X}$ such that $x \in B$. So, $y = f(x) \in f[B]$. Therefore, $y \in \bigcup\{f[A] \mid A \in \boldsymbol{X}\}$. Since $y \in f[\bigcup \boldsymbol{X}]$ was arbitrary, we see that $f[\bigcup \boldsymbol{X}] \subseteq \bigcup\{f[A] \mid A \in \boldsymbol{X}\}$.

Now, let $y \in \bigcup\{f[A] \mid A \in X\}$. Then there is $B \in X$ such that $y \in f[B]$. So, there is $x \in B$ such that $y = f(x)$. By Problem 9 (part (i)) from Problem Set 2, $B \subseteq \bigcup \boldsymbol{X}$. Since $x \in B$ and $B \subseteq \bigcup \boldsymbol{X}$, $x \in \bigcup \boldsymbol{X}$. Thus, $y = f(x) \in f[\bigcup \boldsymbol{X}]$. Since $y \in \bigcup\{f[A] \mid A \in \boldsymbol{X}\}$ was arbitrary, we see that $\bigcup\{f[A] \mid A \in \boldsymbol{X}\} \subseteq f[\bigcup \boldsymbol{X}]$.

Since $f[\bigcup \boldsymbol{X}] \subseteq \bigcup\{f[A] \mid A \in \boldsymbol{X}\}$ and $\bigcup\{f[A] \mid A \in X\} \subseteq f[\bigcup \boldsymbol{X}]$, it follows that $f[\bigcup \boldsymbol{X}] = \bigcup\{f[A] \mid A \in \boldsymbol{X}\}$. □

(ii) Let $y \in f[\bigcap \boldsymbol{X}]$. Then there is $x \in \bigcap \boldsymbol{X}$ such that $f(x) = y$. Let $B \in \boldsymbol{X}$. Since $x \in \bigcap \boldsymbol{X}$, $x \in B$. So, $y = f(x) \in f[B]$. Since $B \in \boldsymbol{X}$ was arbitrary, $y \in \bigcap\{f[A] \mid A \in \boldsymbol{X}\}$. Since $y \in f[\bigcap \boldsymbol{X}]$ was arbitrary, we see that $f[\bigcap \boldsymbol{X}] \subseteq \bigcap\{f[A] \mid A \in \boldsymbol{X}\}$. □

(iii) $x \in f^{-1}[\bigcup \boldsymbol{X}]$ if and only if $f(x) \in \bigcup \boldsymbol{X}$ if and only if there is $A \in \boldsymbol{X}$ such that $f(x) \in A$ if and only if there is $A \in \boldsymbol{X}$ such that $x \in f^{-1}[A]$ if and only if $x \in \bigcup\{f^{-1}[A] \mid A \in \boldsymbol{X}\}$. Therefore, $f^{-1}[\bigcup \boldsymbol{X}] = \bigcup\{f^{-1}[A] \mid A \in \boldsymbol{X}\}$. □

(iv) $x \in f^{-1}[\bigcap \boldsymbol{X}]$ if and only if $f(x) \in \bigcap \boldsymbol{X}$ if and only for all $A \in \boldsymbol{X}$, $f(x) \in A$ if and only if for all $A \in \boldsymbol{X}$, $x \in f^{-1}[A]$ if and only if $x \in \bigcap\{f^{-1}[A] \mid A \in \boldsymbol{X}\}$. Therefore, we see that $f^{-1}[\bigcap \boldsymbol{X}] = \bigcap\{f^{-1}[A] \mid A \in \boldsymbol{X}\}$. □

16. Prove that for any sets A, B, and C, ${}^{B\times C}A \sim {}^{C}({}^{B}A)$.

Proof: Let A, B, and C be sets, and define $F\colon {}^{B\times C}A \to {}^{C}({}^{B}A)$ by $F(f)(c)(b) = f(b,c)$.

Suppose $f, g \in {}^{B\times C}A$ with $F(f) = F(g)$. Let $c \in C$. Since $F(f) = F(g)$, $F(f)(c) = F(g)(c)$. So, for all $b \in B$, $F(f)(c)(b) = F(g)(c)(b)$. So, for all $b \in B$, $f(b,c) = g(b,c)$. Since $c \in C$ was arbitrary, for all $b \in B$ and $c \in C$, $f(b,c) = g(b,c)$. Therefore, $f = g$. Since $f, g \in {}^{B\times C}A$ were arbitrary, F is injective.

Let $k \in {}^{C}({}^{B}A)$ and define $f \in {}^{B\times C}A$ by $f(b,c) = k(c)(b)$. Then $F(f)(c)(b) = f(b,c) = k(c)(b)$. So, $F(f) = k$. Since $k \in {}^{C}({}^{B}A)$ was arbitrary, F is surjective.

Since F is injective and surjective, ${}^{B\times C}A \sim {}^{C}({}^{B}A)$. □

17. Prove the following:

(i) $\mathcal{P}(\mathbb{N}) \sim \{f \in {}^{\mathbb{N}}\mathbb{N} \mid f \text{ is a bijection}\}$.

(ii) ${}^{\mathbb{N}}\mathbb{R} \nsim {}^{\mathbb{R}}\mathbb{N}$.

Proofs:

(i) Let $S = \{f \in {}^{\mathbb{N}}\mathbb{N} \mid f \text{ is a bijection}\}$. Then $S \subseteq {}^{\mathbb{N}}\mathbb{N}$. So $S \preccurlyeq {}^{\mathbb{N}}\mathbb{N}$ by Note 1 following Example 4.16. By part (iv) of Problem 13, ${}^{\mathbb{N}}\mathbb{N} \sim \mathcal{P}(\mathbb{N})$. So, ${}^{\mathbb{N}}\mathbb{N} \preccurlyeq \mathcal{P}(\mathbb{N})$. By the transitivity of $\preccurlyeq$, $S \preccurlyeq \mathcal{P}(\mathbb{N})$.

Now, define $F: \mathcal{P}(\mathbb{N}) \to S$ by $F(A) = f_A$, where f_A is defined as follows: if $n \notin A$, then $f_A(2n) = 2n$ and $f_A(2n+1) = 2n+1$; if $n \in A$, then $f_A(2n) = 2n+1$ and $f_A(2n+1) = 2n$.

To see that F is injective, suppose that $A, B \in \mathcal{P}(\mathbb{N})$ and $A \neq B$. Without loss of generality, suppose that there is $n \in A \setminus B$. Then $f_A(2n) = 2n+1$ and $f_B(2n) = 2n$. So, $f_A \neq f_B$. Thus, $F(A) \neq F(B)$, and therefore, F is injective.

Since $S \preccurlyeq \mathcal{P}(\mathbb{N})$ and $\mathcal{P}(\mathbb{N}) \preccurlyeq S$, by the Cantor-Schroeder-Bernstein Theorem, $\mathcal{P}(\mathbb{N}) \sim S$. □

(ii) By part (iii) of Problem 13, $\mathbb{R} \sim [0,1)$. By Example 4.17, $[0,1) \sim \mathcal{P}(\mathbb{N})$. So, by the transitivity of $\sim$, $\mathbb{R} \sim \mathcal{P}(\mathbb{N})$. By Problem 14, ${}^{\mathbb{N}}\mathbb{R} \sim {}^{\mathbb{N}}\mathcal{P}(\mathbb{N})$.

Using previous equinumerosity results, we get the following:

${}^{\mathbb{N}}\mathbb{R} \sim {}^{\mathbb{N}}\mathcal{P}(\mathbb{N}) \sim {}^{\mathbb{N}}({}^{\mathbb{N}}2) \sim {}^{\mathbb{N}\times\mathbb{N}}2 \sim {}^{\mathbb{N}}2 \sim \mathcal{P}(\mathbb{N}) \sim \mathbb{R} \prec \mathcal{P}(\mathbb{R}) \sim {}^{\mathbb{R}}2 \subseteq {}^{\mathbb{R}}\mathbb{N}$. It follows that ${}^{\mathbb{N}}\mathbb{R} \prec {}^{\mathbb{R}}\mathbb{N}$. □

Note: To help us understand the function F defined in part (i) above, let's draw a visual representation of $F(\mathbb{E})$, where $\mathbb{E}$ is the set of even natural numbers.

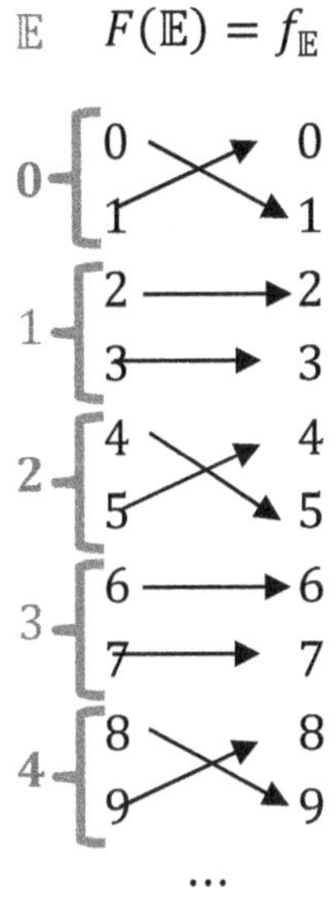

Along the left of the image we have listed the natural numbers $0, 1, 2, 3, 4, \ldots$ (we stopped at 4, but our intention is that they keep going). The elements of $\mathbb{E}$ are $0, 2, 4, \ldots$ We highlighted these in bold. We associate each natural number n with the pair $\{2n, 2n+1\}$. For example, $2 \cdot 4 = 8$ and $2 \cdot 4 + 1 = 9$. So, we associate 4 with the pair of natural numbers $\{8, 9\}$. We used left braces to indicate that association. The arrows give a visual representation of $f_{\mathbb{E}}$. Since $0 \in \mathbb{E}$, $f_{\mathbb{E}}$ swaps the corresponding pair 0 and 1. Since $1 \notin \mathbb{E}$, $f_{\mathbb{E}}$ leaves the corresponding pair 2 and 3 fixed. And so on, down the line...

The configuration of $f_{\mathbb{O}}$, where $\mathbb{O}$ is the set of odd natural numbers would be the opposite of the configuration for the evens. For example, 0 and 1 would remain fixed, while 2 and 3 would be swapped.

Problem Set 5

LEVEL 1

1. Use the Principle of Mathematical Induction to prove each of the following:

   (i) $2^n > n$ for all natural numbers $n \geq 1$.

   (ii) $0 + 1 + 2 + \cdots + n = \frac{n(n+1)}{2}$ for all natural numbers.

   (iii) $n! > 2^n$ for all natural numbers $n \geq 4$ (where $n! = 1 \cdot 2 \cdots n$ for all natural numbers $n \geq 1$).

   (iv) $2^n \geq n^2$ for all natural numbers $n \geq 4$.

Proofs:

(i) **Base Case** ($k = 1$): $2^1 = 2 > 1$.

Inductive Step: Let $k \in \mathbb{N}$ with $k \geq 1$ and assume that $2^k > k$. Then we have

$$2^{k+1} = 2^k \cdot 2^1 = 2^k \cdot 2 > k \cdot 2 = 2k = k + k \geq k + 1.$$

Therefore, $2^{k+1} > k + 1$.

By the Principle of Mathematical Induction, $2^n > n$ for all natural numbers $n \geq 1$. □

(ii) **Base Case** ($k = 0$): $0 = \frac{0(0+1)}{2}$.

Inductive Step: Let $k \in \mathbb{N}$ and assume that $0 + 1 + 2 + \cdots + k = \frac{k(k+1)}{2}$. Then we have

$$0 + 1 + 2 + \cdots + k + (k+1) = \frac{k(k+1)}{2} + (k+1) = (k+1)\left(\frac{k}{2} + 1\right) = (k+1)\left(\frac{k}{2} + \frac{2}{2}\right)$$

$$= (k+1)\left(\frac{k+2}{2}\right) = \frac{(k+1)(k+2)}{2} = \frac{(k+1)\big((k+1)+1\big)}{2}$$

By the Principle of Mathematical Induction, $0 + 1 + 2 + \cdots + n = \frac{n(n+1)}{2}$ for all natural numbers n. □

(iii) **Base Case** ($k = 4$): $4! = 1 \cdot 2 \cdot 3 \cdot 4 = 24 > 16 = 2^4$.

Inductive Step: Let $k \in \mathbb{N}$ with $k \geq 4$ and assume that $k! > 2^k$. Then we have

$$(k+1)! = (k+1)k! > (k+1)2^k \geq (4+1) \cdot 2^k = 5 \cdot 2^k \geq 2 \cdot 2^k = 2^1 \cdot 2^k = 2^{1+k} = 2^{k+1}.$$

Therefore, $(k+1)! > 2^{k+1}$.

By the Principle of Mathematical Induction, $n! > 2^n$ for all natural numbers $n \geq 4$. □

(iv) **Base Case** ($k = 4$): $2^4 = 16 = 4^2$. So, $2^4 \geq 4^2$.

Inductive Step: Let $k \in \mathbb{N}$ with $k \geq 4$ and assume that $2^k \geq k^2$. Then we have

$$2^{k+1} = 2^k \cdot 2^1 \geq k^2 \cdot 2 = 2k^2 = k^2 + k^2.$$

By Theorem 5.14, $k^2 > 2k + 1$. So, we have $2^{k+1} > k^2 + 2k + 1 = (k + 1)^2$.

Therefore, $2^{k+1} \geq (k + 1)^2$.

By the Principle of Mathematical Induction, $2^n \geq n^2$ for all $n \in \mathbb{N}$ with $n \geq 4$. □

Note: Let's take one last look at number (iv). $2^0 = 1 \geq 0 = 0^2$. So, the statement in (iv) is true for $k = 0$. Also, $2^1 = 2 \geq 1 = 1^2$ and $2^2 = 4 = 2^2$. So, the statement is true for $k = 1$ and $k = 2$. However, $2^3 = 8$ and $3^2 = 9$. So, the statement is false for $k = 3$. It follows that $2^n \geq n^2$ for all natural numbers n except $n = 3$.

2. A natural number n is **divisible** by a natural number k, written $k|n$, if there is another natural number b such that $n = kb$. Prove that $n^3 - n$ is divisible by 3 for all natural numbers n.

Proof by Mathematical Induction:

Base Case ($k = 0$): $0^3 - 0 = 0 = 3 \cdot 0$. So, $0^3 - 0$ is divisible by 3.

Inductive Step: Let $k \in \mathbb{N}$ and assume that $k^3 - k$ is divisible by 3. Then $k^3 - k = 3b$ for some integer b. Now,

$$(k + 1)^3 - (k + 1) = (k + 1)[(k + 1)^2 - 1] = (k + 1)[(k + 1)(k + 1) - 1]$$
$$= (k + 1)(k^2 + 2k + 1 - 1) = (k + 1)(k^2 + 2k) = k^3 + 2k^2 + k^2 + 2k = k^3 + 3k^2 + 2k$$
$$= k^3 - k + k + 3k^2 + 2k = (k^3 - k) + 3k^2 + 3k = 3b + 3(k^2 + k) = 3(b + k^2 + k).$$

Since $\mathbb{Z}$ is closed under addition and multiplication, $b + k^2 + k \in \mathbb{Z}$. Therefore, $(k + 1)^3 - (k + 1)$ is divisible by 3.

By the Principle of Mathematical Induction, $n^3 - n$ is divisible by 3 for all $n \in \mathbb{N}$. □

Note: Notice our use of SACT (see Note 7 following the proof of Theorem 5.11) in the beginning of the last line of the sequence of equations. We needed $k^3 - k$ to appear, but the $-k$ was nowhere to be found. So, we simply threw it in, and then repaired the damage by adding k right after it.

3. Let $z = -4 - i$ and $w = 3 - 5i$. Compute each of the following:
   (i) $z + w$
   (ii) zw
   (iii) $\text{Im}\, w$

Solutions:

(i) $z + w = (-4 - i) + (3 - 5i) = (-4 + 3) + (-1 - 5)i = \mathbf{-1 - 6i}$.

(ii) $zw = (-4 - i)(3 - 5i) = (-12 - 5) + (20 - 3)i = \mathbf{-17 + 17i}$.

(iii) $\text{Im}\, w = \text{Im}\,(3 - 5i) = \mathbf{-5}$.

LEVEL 2

4. Prove each of the following. (You may assume that $<$ is a strict linear ordering of $\mathbb{N}$.)
   (i) Addition is commutative in $\mathbb{N}$.
   (ii) The set of natural numbers is closed under multiplication.
   (iii) 1 is a multiplicative identity in $\mathbb{N}$.
   (iv) Multiplication is distributive over addition in $\mathbb{N}$.
   (v) Multiplication is associative in $\mathbb{N}$.
   (vi) Multiplication is commutative in $\mathbb{N}$.
   (vii) For all natural numbers m, n, and k, if $m + k = n + k$, then $m = n$.
   (viii) For all natural numbers m, n, and k, if $mk = nk$, then $m = n$.
   (ix) For all natural numbers m and n, $m < n$ if and only if there is a natural number $k > 0$ such that $n = m + k$.
   (x) For all natural numbers m, n, and k, $m < n$ if and only if $m + k < n + k$.
   (xi) For all natural numbers m and n, if $m > 0$ and $n > 0$, then $mn > 0$.

Proofs:

(i) We first prove by induction on n that $1 + n = n + 1$.

Base Case ($k = 0$): By definition of addition of natural numbers, we have $1 + 0 = 1$. By Theorem 5.8, we have $0 + 1 = 1$. Therefore, $1 + 0 = 0 + 1$.

Inductive Step: Let $k \in \mathbb{N}$ and assume that $1 + k = k + 1$. Then we have

$$1 + (k + 1) = (1 + k) + 1 = (k + 1) + 1.$$

For the first equality, we used the definition of addition of natural numbers. For the second equality, we used the inductive hypothesis.

By the Principle of Mathematical Induction, for all natural numbers n, $1 + n = n + 1$.

We are now ready to use induction to prove the result. Assume that m is a natural number.

Base Case ($k = 0$): By definition of addition of natural numbers, $m + 0 = m$. By Theorem 5.8, $0 + m = m$. Therefore, $m + 0 = 0 + m$.

Inductive Step: Let $k \in \mathbb{N}$ and assume that $m + k = k + m$. Then we have

$$m + (k + 1) = (m + k) + 1 = (k + m) + 1 = k + (m + 1) = k + (1 + m) = (k + 1) + m.$$

For the first and third equalities, we used the definition of addition of natural numbers (or Theorem 5.9). For the second equality, we used the inductive hypothesis. For the fourth equality, we used the preliminary result that we proved above. For the fifth equality, we used Theorem 5.9.

By the Principle of Mathematical Induction, for all natural numbers n, $m + n = n + m$.

Since m was an arbitrary natural number, we have shown that for all natural numbers m and n, we have $m+n=n+m$. □

(ii) Assume that m is a natural number.

Base Case ($k=0$)**:** $m \cdot 0 = 0$, which is a natural number.

Inductive Step: Let k be a natural number and assume that mk is also a natural number. Then $m(k+1) = mk + m$. Since mk and m are both natural numbers, by Theorem 5.7, $mk + m$ is a natural number.

By the Principle of Mathematical Induction, mn is a natural number for all natural numbers n.

Since m was an arbitrary natural number, we have shown that the product of any two natural numbers is a natural number. □

(iii) Assume that m is a natural number.

We have $m \cdot 1 = m(0+1) = m \cdot 0 + m = 0 + m = m$. For the first and fourth equalities, we used Theorem 5.8. For the second and third equalities, we used the definition of multiplication of natural numbers.

We prove that $1 \cdot n = n$ by induction on n.

Proof: Base Case ($k=0$)**:** $1 \cdot 0 = 0$ by the definition of multiplication of natural numbers.

Inductive Step: Let $k \in \mathbb{N}$ and assume that $1 \cdot k = k$. Then

$$1(k+1) = 1 \cdot k + 1 = k + 1.$$

For the first equality, we used the definition of multiplication of natural numbers. For the second equality, we used the inductive hypothesis.

By the Principle of Mathematical Induction, for all natural numbers n, $1 \cdot n = n$. □

(iv) Let m and n be natural numbers. We first prove that for all $t \in \mathbb{N}$, $(m+n) \cdot t = mt + nt$ (we say that multiplication is **right distributive** over addition in $\mathbb{N}$).

Base Case ($k=0$)**:** $(m+n) \cdot 0 = 0$ by the definition of multiplication of natural numbers. Similarly, $m \cdot 0 = 0$ and $n \cdot 0 = 0$. So, $m \cdot 0 + n \cdot 0 = 0$. Therefore, $(m+n) \cdot 0 = m \cdot 0 + n \cdot 0$.

Inductive Step: Let $k \in \mathbb{N}$ and assume that $(m+n) \cdot k = mk + nk$. Then

$$(m+n)(k+1) = (m+n) \cdot k + (m+n) = (mk+nk) + (m+n)$$
$$= (mk+m) + (nk+n) = m(k+1) + n(k+1).$$

For the first and fourth equalities, we used the definition of multiplication of natural numbers. For the second equality, we used the inductive hypothesis. For the third equality we used the fact that addition is associative and commutative in $\mathbb{N}$ several times.

By the Principle of Mathematical Induction, for all natural numbers t, $(m+n) \cdot t = mt + nt$.

We next prove that $n \cdot 0 = 0$ and $0 \cdot n = 0$ for all natural numbers n.

$n \cdot 0 = 0$ by the definition of multiplication of natural numbers.

We prove that $0 \cdot n = 0$ by induction on n.

Base Case ($k = 0$)**:** By definition of multiplication of natural numbers, we have $0 \cdot 0 = 0$.

Inductive Step: Let $k \in \mathbb{N}$ and assume that $0 \cdot k = 0$. Then we have

$$0 \cdot (k+1) = 0 \cdot k + 0 = 0 + 0 = 0.$$

For the first equality, we used the definition of multiplication of natural numbers. For the second equality, we used the inductive hypothesis. For the third equality, we used the definition of addition of natural numbers.

By the Principle of Mathematical Induction, for all natural numbers n, $0 \cdot n = n$.

Let m be a natural number. We prove that for all $n \in \mathbb{N}$, $mn = nm$ (we say that multiplication is **commutative** in $\mathbb{N}$).

Base Case ($k = 0$)**:** $m \cdot 0 = 0$ by the definition of multiplication of natural numbers. We just proved that $0 \cdot m = 0$. Therefore, $m \cdot 0 = 0 \cdot m$

Inductive Step: Let $k \in \mathbb{N}$ and assume that $mk = km$. Then

$$m(k+1) = mk + m = km + m = (k+1)m.$$

For the first equality, we used the definition of multiplication of natural numbers. For the second equality, we used the inductive hypothesis. For the third equality we used the fact that multiplication is right distributive over addition in $\mathbb{N}$ (proved above).

By the Principle of Mathematical Induction, for all natural numbers n, $mn = nm$.

Finally, let $m, n, t \in \mathbb{N}$. Then $m(n+t) = (n+t)m = nm + tm = mn + mt$. This shows that multiplication is distributive over addition in $\mathbb{N}$. □

(v) Assume that m and n are natural numbers

Base Case ($k = 0$)**:** $(mn) \cdot 0 = 0 = m \cdot 0 = m(n \cdot 0)$ by the definition of multiplication of natural numbers.

Inductive Step: Let $k \in \mathbb{N}$ and assume that $(mn)k = m(nk)$. Then

$$(mn)(k+1) = (mn)k + mn = m(nk) + mn = mn + m(nk)$$
$$= m(n + nk) = m(nk + n) = m\big(n(k+1)\big).$$

For the first and sixth equalities, we used the definition of multiplication of natural numbers. For the second equality, we used the inductive hypothesis. For the third and fifth equalities, we used the fact that addition is commutative in $\mathbb{N}$. For the fourth equality, we used the fact that multiplication is distributive over addition in $\mathbb{N}$.

By the Principle of Mathematical Induction, for all natural numbers t, $(mn)t = m(nt)$. This shows that multiplication is associative in $\mathbb{N}$. □

(vi) This was already proved in (iv) above. □

(vii) Let $m, n \in \mathbb{N}$. We prove by induction on k that $m + k = n + k \to m = n$.

Base Case ($k = 0$ and $k = 1$): If $m + 0 = n + 0$, then since $m + 0 = m$ and $n + 0 = n$ (by definition of addition of natural numbers), $m = n$. Next, suppose $m + 1 = n + 1$. Then $m \cup \{m\} = n \cup \{n\}$. If $n \neq m$, then either $n \in m$ or $m \in n$. Without loss of generality, assume that $n \in m$. Since $\in$ is antisymmetric on $\mathbb{N}$ and $n \neq m$, we must have $m \notin n$. Thus, $m = n$, contrary to our assumption that $n \neq m$. This contradiction shows that $m = n$.

Inductive Step: Let $t \in \mathbb{N}$, assume that $m + t = n + t \to m = n$, and let $m + (t + 1) = n + (t + 1)$. By the definition of addition in $\mathbb{N}$, $(m + t) + 1 = (n + t) + 1$. By the base case, $m + t = n + t$. By the inductive hypothesis, $m = n$.

By the Principle of Mathematical Induction, for all natural numbers m, n, and k, if $m + k = n + k$, then $m = n$. □

(viii) Let $m, n \in \mathbb{N}$. We prove by induction on k that $mk = nk \to m = n$.

Base Case ($k = 0$): If $m \cdot 0 = n \cdot 0$, then since $m \cdot 0 = 0$ and $n \cdot 0 = 0$ (by definition of multiplication of natural numbers), $m = n$.

Inductive Step: Let $t \in \mathbb{N}$, assume that $mt = nt \to m = n$, and let $m(t + 1) = n(t + 1)$. By the definition of multiplication in $\mathbb{N}$, $mt + 1 = nt + 1$. By (vii), $mt = nt$. By the inductive hypothesis, we have $m = n$.

By the Principle of Mathematical Induction, for all natural numbers m, n, and k, if $mk = nk$, then $m = n$. □

(ix) Let $m \in \mathbb{N}$. We prove by induction on n that if $m < n$, there is $k > 0$ such that $n = m + k$.

Base Case ($t = 0$): If $m < 0$, then $m \in \emptyset$, which is impossible. So, the conclusion is vacuously true.

Inductive Step: Let $t \in \mathbb{N}$ and assume that if $m < t$, there is $k > 0$ such that $t = m + k$. Assume that $m < t + 1$. Then $m < t$ or $m = t$. If $m < t$, then $t + 1 = (m + k) + 1 = m + (k + 1)$. If $m = t$, then $t + 1 = m + 1$.

By the Principle of Mathematical Induction, for all $n \in \mathbb{N}$, if $m < n$, there is $k > 0$ so that $n = m + k$.

Now, let $n, m \in \mathbb{N}$. We prove by induction on $k > 0$ that if $n = m + k$, then $m < n$.

Base Case ($t = 1$): If $n = m + 1 = m \cup \{m\}$, then since $m \in \{m\}$, $m \in n$, and so, $m < n$.

Inductive Step: Assume that if $n = m + t$, then $m < n$. Let $n = m + (t + 1)$. Then since addition is commutative and associative in $\mathbb{N}$, $n = m + (1 + t) = (m + 1) + t$. By the inductive hypothesis, we have $m + 1 < n$. Since $m < m + 1$ and $<$ is transitive on $\mathbb{N}$, $m < n$.

By the Principle of Mathematical Induction, if $k > 0$ and $n = m + k$, then $m < n$. □

(x) Let $m, n, k \in \mathbb{N}$.

By (ix), $m < n$ if and only if there is a natural number $t > 0$ such that $n = m + t$. Now, $n + k = (m + t) + k = m + (t + k) = m + (k + t) = (m + k) + t$. So, if $m < n$, then there is a natural number $t > 0$ such that $n + k = (m + k) + t$. Thus, by (ix), $m + k < n + k$. Conversely, if $m + k < n + k$, then there is a natural number t such that $n + k = (m + k) + t = (m + t) + k$. By (vii), $n = m + t$. So, by (ix) again, $m < n$. □

(xi) Let $m \in \mathbb{N}$ with $m > 0$. We prove by induction on n that $n > 0 \to mn > 0$.

Base Case ($k = 1$): If $n = 1$, then $m \cdot 1 = m \cdot 0 + m = 0 + m = m > 0$.

Inductive Step: Let $k \in \mathbb{N}$ with $k > 0$ and assume that $mk > 0$. Then $m(k+1) = mk + k$. Since $mk > 0$, by (x), $mk + k > 0 + k = k > 0$. So, $m(k+1) > 0$.

By the Principle of Mathematical Induction, for all natural numbers m and n, if $m > 0$ and $n > 0$, then $mn > 0$. □

5. A set A is **transitive** if $\forall x(x \in A \to x \subseteq A)$ (in words, every element of A is also a subset of A). Prove that every natural number is transitive.

Proof by Mathematical Induction:

Base Case ($k = 0$): $0 = \emptyset$. Since $\emptyset$ has no elements, it is vacuously true that every element of $\emptyset$ is a subset of $\emptyset$.

Inductive Step: Assuming that k is transitive, let $j \in k + 1 = k \cup \{k\}$ and $m \in j$. Then $j \in k$ or $j \in \{k\}$. If $j \in k$, then we have $m \in j \in k$. Since k is transitive, $m \in k$. Therefore, $m \in k \cup \{k\} = k + 1$. If $j \in \{k\}$, then $j = k$. So, $m \in k$, and again, $m \in k \cup \{k\} = k + 1$.

By the Principle of Mathematical Induction, every natural number is transitive. □

6. Determine if each of the following sequences are Cauchy sequences. Are any of the Cauchy sequences equivalent?

   (i) $(x_n) = \left(1 + \frac{1}{n+1}\right)$

   (ii) $(y_n) = (2^n)$

   (iii) $(z_n) = \left(1 - \frac{1}{2n+1}\right)$

Solutions:

(i) **Cauchy**

(ii) **Not Cauchy**

(iii) **Cauchy**

(x_n) and (z_n) are equivalent.

LEVEL 3

7. Each of the following complex numbers are written in exponential form. Rewrite each complex number in standard form: (i) $e^{\pi i}$; (ii) $e^{-\frac{5\pi}{2}i}$; (iii) $3e^{\frac{\pi}{4}i}$; (iv) $2e^{\frac{\pi}{3}i}$; (v) $\sqrt{2}e^{\frac{7\pi}{6}i}$; (vi) $\pi e^{-\frac{5\pi}{4}i}$; (vii) $e^{\frac{19\pi}{12}}$

Solutions:

(i) $e^{\pi i} = \cos\pi + i\sin\pi = -1 + 0i = \mathbf{-1}.$

(ii) $e^{-\frac{5\pi}{2}i} = \cos\left(-\frac{5\pi}{2}\right) + i\sin\left(-\frac{5\pi}{2}\right) = \cos\frac{5\pi}{2} - i\sin\frac{5\pi}{2} = 0 - 1i = \boldsymbol{-i}.$

(iii) $3e^{\frac{\pi}{4}i} = 3\left(\cos\frac{\pi}{4} + i\sin\frac{\pi}{4}\right) = 3\left(\frac{\sqrt{2}}{2} + \frac{\sqrt{2}}{2}i\right) = \mathbf{\frac{3\sqrt{2}}{2}} + \boldsymbol{\frac{3\sqrt{2}}{2}i}.$

(iv) $2e^{\frac{\pi}{3}i} = 2\left(\cos\frac{\pi}{3} + i\sin\frac{\pi}{3}\right) = 2\left(\frac{1}{2} + \frac{\sqrt{3}}{2}i\right) = \mathbf{1} + \boldsymbol{\sqrt{3}i}.$

(v) $\sqrt{2}e^{\frac{7\pi}{6}i} = \sqrt{2}\left(\cos\frac{7\pi}{6} + i\sin\frac{7\pi}{6}\right) = \sqrt{2}\left(-\frac{\sqrt{3}}{2} - \frac{1}{2}i\right) = \mathbf{-\frac{\sqrt{6}}{2}} - \boldsymbol{\frac{\sqrt{2}}{2}i}.$

(vi) $\pi e^{-\frac{5\pi}{4}i} = \pi\left(\cos\left(-\frac{5\pi}{4}\right) + i\sin\left(-\frac{5\pi}{4}\right)\right) = \pi\left(\cos\frac{5\pi}{4} - i\sin\frac{5\pi}{4}\right) = \mathbf{-\frac{\pi\sqrt{2}}{2}} + \boldsymbol{\frac{\pi\sqrt{2}}{2}i}.$

(vii) $e^{\frac{19\pi}{12}} = \cos\frac{19\pi}{12} + i\sin\frac{19\pi}{12} = \mathbf{\frac{-\sqrt{2}+\sqrt{6}}{4}} + \boldsymbol{\frac{-\sqrt{2}-\sqrt{6}}{4}i}.$

8. Each of the following complex numbers are written in standard form. Rewrite each complex number in exponential form: (i) $-1 - i$; (ii) $\sqrt{3} + i$; (iii) $1 - \sqrt{3}i$; (iv) $\left(\frac{\sqrt{6}+\sqrt{2}}{4}\right) + \left(\frac{\sqrt{6}-\sqrt{2}}{4}\right)i$.

Solutions:

(i) $r^2 = (-1)^2 + (-1)^2 = 1 + 1 = 2$. So, $r = \sqrt{2}$. $\tan\theta = \frac{-1}{-1} = 1$. So, $\theta = \pi + \frac{\pi}{4} = \frac{5\pi}{4}$. Therefore, $-1 - i = \sqrt{2}e^{\frac{5\pi}{4}i} = \boldsymbol{\sqrt{2}e^{-\frac{3\pi}{4}i}}$.

(ii) $r^2 = \left(\sqrt{3}\right)^2 + 1^2 = 3 + 1 = 4$. So, $r = 2$. $\tan\theta = \frac{1}{\sqrt{3}}$. So, $\theta = \frac{\pi}{6}$. Therefore, we have $\sqrt{3} + i = \boldsymbol{2e^{\frac{\pi}{6}i}}$.

(iii) $r^2 = 1^2 + \left(-\sqrt{3}\right)^2 = 1 + 3 = 4$. So, $r = 2$. $\tan\theta = \frac{-\sqrt{3}}{1}$. So, $\theta = -\frac{\pi}{3}$. Therefore, we have $1 - \sqrt{3}i = \boldsymbol{2e^{-\frac{\pi}{3}i}}$.

(iv) $r^2 = \left(\frac{\sqrt{6}+\sqrt{2}}{4}\right)^2 + \left(\frac{\sqrt{6}-\sqrt{2}}{4}\right)^2 = \frac{6+2+2\sqrt{12}}{16} + \frac{6+2-2\sqrt{12}}{16} = \frac{16}{16} = 1$. So, $r = 1$.

By Theorem 5.23, we have the following:

$$\cos\frac{\pi}{12} = \cos\left(\frac{\pi}{4} - \frac{\pi}{6}\right) = \cos\frac{\pi}{4}\cos\frac{\pi}{6} + \sin\frac{\pi}{4}\sin\frac{\pi}{6} = \frac{\sqrt{2}}{2}\cdot\frac{\sqrt{3}}{2} + \frac{\sqrt{2}}{2}\cdot\frac{1}{2} = \mathbf{\frac{\sqrt{6}+\sqrt{2}}{4}}.$$

$$\sin\frac{\pi}{12} = \sin\left(\frac{\pi}{4} - \frac{\pi}{6}\right) = \sin\frac{\pi}{4}\cos\frac{\pi}{6} - \cos\frac{\pi}{4}\sin\frac{\pi}{6} = \frac{\sqrt{2}}{2}\cdot\frac{\sqrt{3}}{2} - \frac{\sqrt{2}}{2}\cdot\frac{1}{2} = \mathbf{\frac{\sqrt{6}-\sqrt{2}}{4}}.$$

Therefore, $\left(\frac{\sqrt{6}+\sqrt{2}}{4}\right) + \left(\frac{\sqrt{6}-\sqrt{2}}{4}\right)i = 1e^{\frac{\pi}{12}i} = \boldsymbol{e^{\frac{\pi}{12}i}}$.

9. Write the following complex numbers in standard form: (i) $\left(\frac{\sqrt{2}}{2} + \frac{\sqrt{2}}{2}i\right)^4$; (ii) $\left(1 + \sqrt{3}i\right)^5$.

Solutions:

(i) If $z = \frac{\sqrt{2}}{2} + \frac{\sqrt{2}}{2}i$, then $r = \sqrt{\left(\frac{\sqrt{2}}{2}\right)^2 + \left(\frac{\sqrt{2}}{2}\right)^2} = \sqrt{\frac{2}{4} + \frac{2}{4}} = 1$ and $\tan\theta = \frac{\frac{\sqrt{2}}{2}}{\frac{\sqrt{2}}{2}} = 1$, so that $\theta = \frac{\pi}{4}$. So, in exponential form, $z = e^{\frac{\pi}{4}i}$. Therefore, $\left(\frac{\sqrt{2}}{2} + \frac{\sqrt{2}}{2}i\right)^4 = \left(e^{\frac{\pi}{4}i}\right)^4 = e^{\pi i} = \mathbf{-1}$.

(ii) If $z = 1 + \sqrt{3}i$, then $r = \sqrt{1^2 + \left(\sqrt{3}\right)^2} = \sqrt{1+3} = \sqrt{4} = 2$ and $\tan\theta = \frac{\sqrt{3}}{1} = \sqrt{3}$, so that $\theta = \frac{\pi}{3}$. So, in exponential form, $z = 2e^{\frac{\pi}{3}i}$. So, $\left(1 + \sqrt{3}i\right)^5 = \left(2e^{\frac{\pi}{3}i}\right)^5 = 2^5 e^{\frac{5\pi}{3}i} = \mathbf{16 - 16\sqrt{3}}\boldsymbol{i}$.

10. Prove that if $n \in \mathbb{N}$ and A is a nonempty subset of n, then A has a least element.

Proof by Mathematical Induction:

Base Case ($k = 0$): $0 = \emptyset$. The only subset of $\emptyset$ is $\emptyset$. So, the statement is vacuously true.

Inductive Step: Assume that every nonempty subset of the natural number k has a least element. We will show that every nonempty subset of $k + 1 = k \cup \{k\}$ has a least element.

Let A be a nonempty subset of $k+1$. Then $A \setminus \{k\} \subseteq k$. If $A \setminus \{k\} \neq \emptyset$, then by the inductive hypothesis, $A \setminus \{k\}$ has a least element, say j. Since $j \in k$, j is the least element of A. If $A \setminus \{k\} = \emptyset$, then k is the only element of A, and therefore, it is the least element of A.

By the Principle of Mathematical Induction, if $n \in \mathbb{N}$ and A is a nonempty subset of n, then A has a least element. □

11. Prove POMI → WOP.

Proof: Assume POMI, let A be a nonempty subset of $\mathbb{N}$, and choose $n \in A$. If $n \cap A = \emptyset$, then n is the least element of A (If $m \in A$ with $m \in n$, then $m \in n \cap A$, contradicting $n \cap A = \emptyset$). Otherwise, $n \cap A$ is a nonempty subset of n, and so, by Problem 5, $n \cap A$ has a least element m. Then m is the least element of A (If $k \in A$ with $k \in m$, then $k \in n$ by Problem 4, and so, m is not the least element of $n \cap A$). □

12. Prove that $<_{\mathbb{Z}}$ is a well-defined strict linear ordering on $\mathbb{Z}$. You may use the fact that $<_{\mathbb{N}}$ is a well-defined strict linear ordering on $\mathbb{N}$.

Proof: We first show that $<_{\mathbb{Z}}$ is well-defined. Suppose that $(a,b) \sim (a',b')$ and $(c,d) \sim (c',d)$. Since $(a,b) \sim (a',b')$, $a + b' = b + a'$. Since $(c,d) \sim (c',d')$, $c + d' = d + c'$.

We need to check that $[(a,b)] <_{\mathbb{Z}} [(c,d)]$ if and only if $[(a',b')] <_{\mathbb{Z}} [(c',d')]$. We have

$[(a,b)] <_{\mathbb{Z}} [(c,d)]$ if and only if $a + d <_{\mathbb{N}} b + c$ if and only if $a + d + b' + c' <_{\mathbb{N}} b + c + b' + c'$ if and only if $a + b' + d + c' <_{\mathbb{N}} b + c + b' + c'$ if and only if $b + a' + c + d' <_{\mathbb{N}} b + c + b' + c'$ if and only if $b + c + a' + d' <_{\mathbb{N}} b + c + b' + c'$ if and only if $a' + d' < b' + c'$ if and only if $[(a',b')] <_{\mathbb{Z}} [(c',d')]$, as desired.

Next, we show that $<_{\mathbb{Z}}$ is antireflexive. To see this, note that $a + b \not<_{\mathbb{N}} a + b$ because $\not<_{\mathbb{N}}$ is antireflexive. So, $a + b \not<_{\mathbb{N}} b + a$. Therefore, $[(a, b)] \not<_{\mathbb{Z}} [(a, b)]$.

To see that $<_{\mathbb{Z}}$ is antisymmetric, suppose that $[(a, b)] <_{\mathbb{Z}} [(c, d)]$ and $[(c, d)] <_{\mathbb{Z}} [(a, b)]$. Then we have $a + d <_{\mathbb{N}} b + c$ and $c + b <_{\mathbb{N}} d + a$, or equivalently, $b + c <_{\mathbb{N}} a + d$. This is impossible, and so, it is vacuously true that $<_{\mathbb{Z}}$ is antisymmetric.

To see that $<_{\mathbb{Z}}$ is transitive, suppose that $[(a, b)] <_{\mathbb{Z}} [(c, d)]$ and $[(c, d)] <_{\mathbb{Z}} [(e, f)]$. Then we have $a + d <_{\mathbb{N}} b + c$ and $c + f <_{\mathbb{N}} d + e$. By adding each side of these two inequalities we get the inequality $a + d + c + f <_{\mathbb{N}} b + c + d + e$. Cancelling c and d from each side of this last inequality yields $a + f <_{\mathbb{N}} b + e$. Therefore, $[(a, b)] <_{\mathbb{Z}} [(e, f)]$.

Finally, we check that trichotomy holds. Suppose $[(a, b)] \not<_{\mathbb{Z}} [(c, d)]$ and $[(a, b)] \neq [(c, d)]$. Then $a + d \not<_{\mathbb{N}} b + c$ and $a + d \neq b + c$. Since trichotomy holds for $<_{\mathbb{N}}$, we have $b + c <_{\mathbb{N}} a + d$, or equivalently, $c + b <_{\mathbb{N}} d + a$. Therefore, $[(c, d)] <_{\mathbb{Z}} [(a, b)]$. □

LEVEL 4

13. Prove that $3^n - 1$ is even for all natural numbers n.

Proof by Mathematical Induction:

Base Case ($k = 0$): $3^0 - 1 = 1 - 1 = 0 = 2 \cdot 0$. So, $3^0 - 1$ is even.

Inductive Step: Let $k \in \mathbb{N}$ and assume that $3^k - 1$ is even. Then $3^k - 1 = 2b$ for some integer b. Now,

$$3^{k+1} - 1 = 3^k \cdot 3^1 - 1 = 3^k \cdot 3 - 1 = 3^k \cdot 3 - 3^k + 3^k - 1 = 3^k(3 - 1) + (3^k - 1)$$
$$= 3^k \cdot 2 + 2b = 2 \cdot 3^k + 2b = 2(3^k + b).$$

Since $\mathbb{N}$ is closed under multiplication, $3^k \in \mathbb{N}$. Since $\mathbb{N}$ is closed under addition, $3^k + b \in \mathbb{N}$. Therefore, $3^{k+1} - 1$ is even.

By the Principle of Mathematical Induction, $3^n - 1$ is even for all $n \in \mathbb{N}$. □

Notes: Notice our use of SACT (see Note 7 following the proof of Theorem 5.11) in the middle of the first line of the sequence of equations. We needed $3^k - 1$ to appear, so we added 3^k, and then subtracted 3^k to the left of it.

14. Show that the Principle of Mathematical Induction is equivalent to the following statement:

(⋆) Let $P(n)$ be a statement and suppose that (i) $P(0)$ is true and (ii) for all $k \in \mathbb{N}$, $P(k) \to P(k + 1)$. Then $P(n)$ is true for all $n \in \mathbb{N}$.

Proof: Recall that the Principle of Mathematical Induction says the following: Let S be a set of natural numbers such that (i) $0 \in S$ and (ii) for all $k \in \mathbb{N}$, $k \in S \to k + 1 \in S$. Then $S = \mathbb{N}$.

Suppose that the Principle of Mathematical Induction is true and let $P(n)$ be a statement such that $P(0)$ is true, and for all $k \in \mathbb{N}$, $P(k) \to P(k+1)$. Define $S = \{n \mid (P(n)\}$. Since $P(0)$ is true, $0 \in S$. If $k \in S$, then $P(k)$ is true. So, $P(k+1)$ is true, and therefore, $k+1 \in S$. By the Principle of Mathematical Induction, $S = \mathbb{N}$. So, $P(n)$ is true for all $n \in \mathbb{N}$.

Now, suppose that $(\star)$ holds, and let S be a set of natural numbers such that $0 \in S$, and for all $k \in \mathbb{N}$, $k \in S \to k+1 \in S$. Let $P(n)$ be the statement $n \in S$. Since $0 \in S$, $P(0)$ is true. If $P(k)$ is true, then $k \in S$. So, $k+1 \in S$, and therefore, $P(k+1)$ is true. By $(\star)$, $P(n)$ is true for all n. So, for all $n \in \mathbb{N}$, we have $n \in S$. In other words, $\mathbb{N} \subseteq S$. Since we were given $S \subseteq \mathbb{N}$, we have $S = \mathbb{N}$. □

15. Prove that addition of integers is well-defined.

Proof: Suppose that $(a,b) \sim (a',b')$ and $(c,d) \sim (c',d)$. Since $(a,b) \sim (a',b')$, $a + b' = b + a'$. Since $(c,d) \sim (c',d')$, $c + d' = d + c'$.

We need to check that $(a+c, b+d) \sim (a'+c', b'+d')$, or equivalently, we need to check that $(a+c)+(b'+d') = (b+d)+(a'+c')$

Since $a + b' = b + a'$ and $c + d' = d + c'$, we have

$$(a+c)+(b'+d') = (a+b')+(c+d') = (b+a')+(d+c') = (b+d)+(a'+c').$$

Therefore, $(a+c, b+d) = (a'+c', b'+d')$, as desired.

16. Prove that addition and multiplication of rational numbers are well-defined.

Proof: Suppose that $\frac{a}{b} = \frac{a'}{b'}$ and $\frac{c}{d} = \frac{c'}{d'}$. Since $\frac{a}{b} = \frac{a'}{b'}$, we have $ab' = ba'$. Since $\frac{c}{d} = \frac{c'}{d'}$, we have $cd' = dc'$.

We first need to check that $\frac{a}{b} + \frac{c}{d} = \frac{a'}{b'} + \frac{c'}{d'}$, or equivalently, $\frac{ad+bc}{bd} = \frac{a'd'+b'c'}{b'd'}$.

Since $ab' = ba'$ and $cd' = dc'$, we have

$$\begin{aligned}(ad+bc)(b'd') &= adb'd' + bcb'd' = ab'dd' + cd'bb' = ba'dd' + dc'bb' \\ &= bda'd' + bdb'c' = (bd)(a'd'+b'c').\end{aligned}$$

Therefore, $\frac{ad+bc}{bd} = \frac{a'd'+b'c'}{b'd'}$, as desired.

We next need to check that $\frac{a}{b} \cdot \frac{c}{d} = \frac{a'}{b'} \cdot \frac{c'}{d'}$, or equivalently, $\frac{ac}{bd} = \frac{a'c'}{b'd'}$.

Since $ab' = ba'$ and $cd' = dc'$, we have

$$(ac)(b'd') = (ab')(cd') = (ba')(dc') = (bd)(a'c')$$

Therefore, $\frac{ac}{bd} = \frac{a'c'}{b'd'}$, as desired. □

17. Let $A = \{(x_n) \mid (x_n)$ is a Cauchy sequence of rational numbers$\}$ and define the relation R on A by $(x_n)R(y_n)$ if and only if for every $k \in \mathbb{N}^+$, there is $K \in \mathbb{N}$ such that $n > K$ implies $|x_n - y_n| < \frac{1}{k}$. Prove that R is an equivalence relation on A.

Proof: Let $(x_n) \in A$, let $k \in \mathbb{N}^+$ and let $K = 0$. Then $n > K$ implies $|x_n - x_n| = 0 < \frac{1}{k}$. So, $(x_n)R(x_n)$, and therefore, R is reflexive.

Since $|x_n - y_n| = |y_n - x_n|$, it is clear that R is symmetric.

Let $(x_n), (y_n), (z_n) \in A$ with $(x_n)R(y_n)$ and $(y_n)R(z_n)$ and let $k \in \mathbb{N}^+$. Since $(x_n)R(y_n)$, there is $K_1 \in \mathbb{N}$ such that $n > K_1$ implies $|x_n - y_n| < \frac{1}{2k}$. Since $(y_n)R(z_n)$, there is $K_2 \in \mathbb{N}$ such that $n > K_2$ implies $|y_n - z_n| < \frac{1}{2k}$. Let $K = \max\{K_1, K_2\}$. Let $n > K$. Since $K \geq K_1$, $n > K_1$, and therefore, we have $|x_n - y_n| < \frac{1}{2k}$. Since $K \geq K_2$, $n > K_2$, and therefore, we have $|y_n - z_n| < \frac{1}{2k}$. It follows that

$$|x_n - z_n| = |x_n - y_n + y_n - z_n| \leq |x_n - y_n| + |y_n - z_n| < \frac{1}{2k} + \frac{1}{2k} = 2 \cdot \frac{1}{2k} = \frac{1}{k}.$$

So, $(x_n)R(z_n)$, and therefore, R is transitive.

Since R is reflexive, symmetric, and transitive, it follows that R is an equivalence relation. □

18. Prove that $\{A \in \mathcal{P}(\mathbb{N}) \mid A \text{ is finite}\}$ is countable and $\{A \in \mathcal{P}(\mathbb{N}) \mid A \text{ is infinite}\}$ is uncountable.

Proof: We first show that $\boldsymbol{X} = \{A \in \mathcal{P}(\mathbb{N}) \mid A \text{ is finite}\}$ is countable. For each $n \in \mathbb{N}$, let $A_n = \{A \in \mathcal{P}(\mathbb{N}) \mid |A| \leq n\}$. Since $\boldsymbol{X} = \bigcup\{A_n \mid n \in \mathbb{N}\}$, by Problem 11 from Problem Set 4, it suffices to show that for each $n \in \mathbb{N}$, A_n is countable. We show this by induction on $n \in \mathbb{N}$. $A_0 = \{\emptyset\}$, which is certainly countable. $\{\{n\} \mid n \in \mathbb{N}\}$ is clearly equinumerous to $\mathbb{N}$ via the function sending $\{n\}$ to n. Therefore, we see that $A_1 = A_0 \cup \{\{n\} \mid n \in \mathbb{N}\}$ is countable. Let $k \in \mathbb{N}$ and assume that A_k is countable. For each $n \in \mathbb{N}$, the set $B_k^n = \{A \cup \{n\} \mid A \in A_k\}$ is countable. By Problem 11 from Problem Set 4, the set $B_{k+1} = \bigcup\{B_k^n \mid n \in \mathbb{N}\}$ is countable. So, $A_{k+1} = A_0 \cup B_{k+1}$ is countable. By the principle of mathematical induction, for each $n \in \mathbb{N}$, A_n is countable. It follows that $\boldsymbol{X} = \{A \in \mathcal{P}(\mathbb{N}) \mid A \text{ is finite}\}$ is countable.

Let $\boldsymbol{Y} = \{A \in \mathcal{P}(\mathbb{N}) \mid A \text{ is infinite}\}$. Since every subset of $\mathbb{N}$ is either finite or infinite, $\mathcal{P}(\mathbb{N}) = \boldsymbol{X} \cup \boldsymbol{Y}$. If $\boldsymbol{Y}$ were countable, then since $\boldsymbol{X}$ is countable, by Problem 11 from Problem Set 4, $\mathcal{P}(\mathbb{N})$ would be countable, which we know it is not. Therefore, $\boldsymbol{Y}$ is uncountable. □

Note: Computing A_1 in the proof above was not necessary. $B_0^n = \{A \cup \{n\} | A \in A_0\} = \{\{n\}\}$. Therefore, $B_1 = \bigcup\{B_0^n \mid n \in \mathbb{N}\} = \{\{n\} \mid n \in \mathbb{N}\}$. So, $A_1 = A_0 \cup B_1 = A_0 \cup \{\{n\} \mid n \in \mathbb{N}\}$. This is the same set that we wrote out explicitly in the proof.

19. Consider triangle AOP, where $O = (0,0)$, $A = (1,0)$, and P is the point on the unit circle so that angle POA has radian measure $\frac{\pi}{3}$. Prove that triangle AOP is equilateral, and then use this to prove that $W\left(\frac{\pi}{3}\right) = \left(\frac{1}{2}, \frac{\sqrt{3}}{2}\right)$. You may use the following facts about triangles: (i) The interior angle measures of a triangle sum to π radians; (ii) Two sides of a triangle have the same length if and only if the interior angles of the triangle opposite these sides have the same measure; (iii) If two sides of a triangle have the same length, then the line segment beginning at the point of intersection of those two sides and terminating on the opposite base midway between the endpoints of that base is perpendicular to that base.

Proof: Let's start by drawing the unit circle together with triangle AOP. We also draw line segment PE, where E is midway between O and A. By (iii), PE is perpendicular to OA.

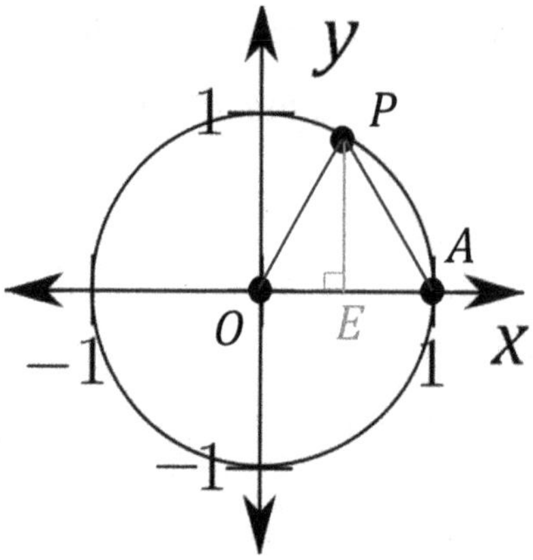

Since OP and OA are both radii of the circle, they have the same length. By (ii), angles OAP and OPA have the same measure. By (i), the sum of these measures is $\pi - \frac{\pi}{3} = \frac{3\pi}{3} - \frac{\pi}{3} = \frac{2\pi}{3}$. So, each of angles OAP and OPA measure $\frac{\pi}{3}$ radians. It follows from (ii) again that triangle AOP is equilateral.

Now, $OP = 1$ because OP is a radius of the unit circle and $OE = \frac{1}{2}$ because OA is a radius of the unit circle and E is midway between O and A. Since triangle OEP is a right triangle with hypotenuse OP, by the Pythagorean Theorem, $PE^2 = OP^2 - OE^2 = 1^2 - \left(\frac{1}{2}\right)^2 = 1 - \frac{1}{4} = \frac{3}{4}$. So, $PE = \sqrt{\frac{3}{4}} = \frac{\sqrt{3}}{\sqrt{4}} = \frac{\sqrt{3}}{2}$. It follows that $W\left(\frac{\pi}{3}\right) = \left(\frac{1}{2}, \frac{\sqrt{3}}{2}\right)$. □

20. Prove that $W\left(\frac{\pi}{6}\right) = \left(\frac{\sqrt{3}}{2}, \frac{1}{2}\right)$. You can use facts (i), (ii), and (iii) described in Problem 19.

Proof: Let's start by drawing a picture similar to what we drew in Problem 19. We draw P and Q on the unit circle and A on the positive x-axis so that angle AOP has radian measure $\frac{\pi}{6}$, angle AOQ has radian measure $-\frac{\pi}{6}$, and A is right in the middle of the line segment joining P and Q.

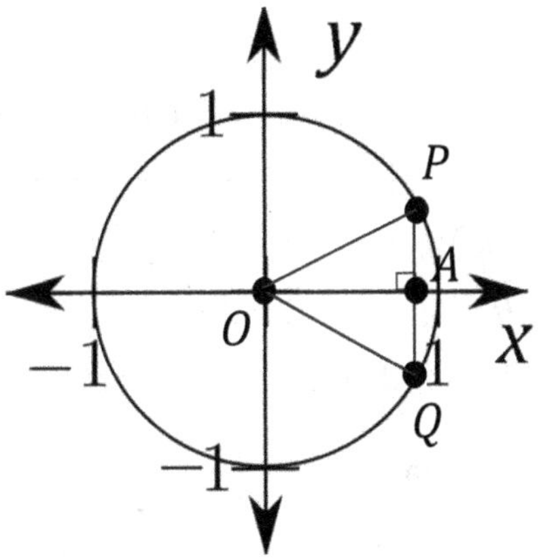

By reasoning similar to what was done in Problem 19, we see that triangle POQ is equilateral and OA is perpendicular to PQ.

Now, $OP = 1$ because OP is a radius of the unit circle and $PA = \frac{1}{2}$ because A is midway between P and Q. Since triangle POA is a right triangle with hypotenuse OP, by the Pythagorean Theorem, $OA^2 = OP^2 - AP^2 = 1^2 - \left(\frac{1}{2}\right)^2 = 1 - \frac{1}{4} = \frac{3}{4}$. Therefore, $OA = \sqrt{\frac{3}{4}} = \frac{\sqrt{3}}{\sqrt{4}} = \frac{\sqrt{3}}{2}$. It follows that $W\left(\frac{\pi}{6}\right) = \left(\frac{\sqrt{3}}{2}, \frac{1}{2}\right)$. □

21. Let θ and ϕ be the radian measure of angles A and B, respectively. Prove the following identity:

$$\cos(\theta - \phi) = \cos\theta\cos\phi + \sin\theta\sin\phi$$

Proof: Let's draw a picture of the unit circle together with angles θ, ϕ, and $\theta - \phi$ in standard position, and label the corresponding points on the unit circle.

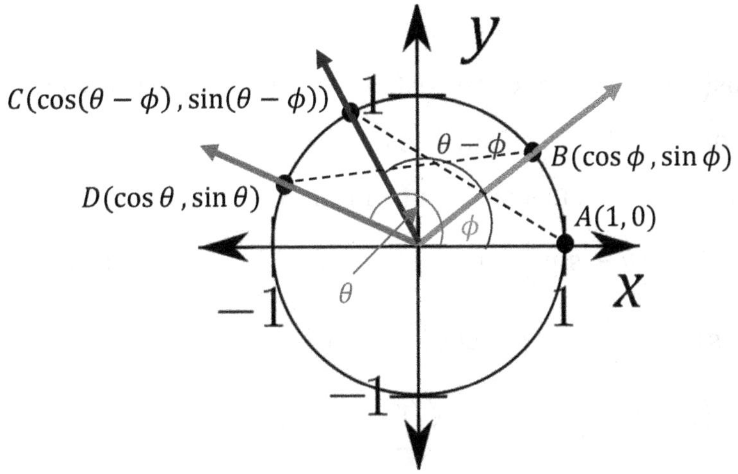

Since the arcs moving counterclockwise from A to C and from B to D both have radian measure $\theta - \phi$, it follows that $AC = BD$, and so, using the Pythagorean Theorem twice, we have

$$(\cos(\theta - \phi) - 1)^2 + (\sin(\theta - \phi) - 0)^2 = (\cos\theta - \cos\phi)^2 + (\sin\theta - \sin\phi)^2$$

The left-hand side of this equation is equal to:

$$(\cos(\theta - \phi) - 1)^2 + (\sin(\theta - \phi) - 0)^2$$
$$= \cos^2(\theta - \phi) - 2\cos(\theta - \phi) + 1 + \sin^2(\theta - \phi)$$
$$= (\cos^2(\theta - \phi) + \sin^2(\theta - \phi)) - 2\cos(\theta - \phi) + 1$$
$$= 1 - 2\cos(\theta - \phi) + 1 \text{ (by the Pythagorean Identity)}$$
$$= 2 - 2\cos(\theta - \phi)$$

The right-hand side of this equation is equal to:

$$(\cos\theta - \cos\phi)^2 + (\sin\theta - \sin\phi)^2$$
$$= \cos^2\theta - 2\cos\theta\cos\phi + \cos^2\phi + \sin^2\theta - 2\sin\theta\sin\phi + \sin^2\phi$$
$$= (\cos^2\theta + \sin^2\theta) + (\cos^2\phi + \sin^2\phi) - 2\cos\theta\cos\phi - 2\sin\theta\sin\phi$$
$$= 1 + 1 - 2\cos\theta\cos\phi - 2\sin\theta\sin\phi$$
$$= 2 - 2\cos\theta\cos\phi - 2\sin\theta\sin\phi$$

Therefore, we have $2 - 2\cos(\theta - \phi) = 2 - 2\cos\theta\cos\phi - 2\sin\theta\sin\phi$. Subtracting 2 from each side of this equation gives us $-2\cos(\theta - \phi) = -2\cos\theta\cos\phi - 2\sin\theta\sin\phi$. Multiplying each side of this last equation by $-\frac{1}{2}$ gives us $\cos(\theta - \phi) = \cos\theta\cos\phi + \sin\theta\sin\phi$, as desired. □

22. Let θ and ϕ be the radian measure of angles A and B, respectively. Prove the following identities: (i) $\cos(\theta + \phi) = \cos\theta\cos\phi - \sin\theta\sin\phi$; (ii) $\cos(\pi - \theta) = -\cos\theta$; (iii) $\cos\left(\frac{\pi}{2} - \theta\right) = \sin\theta$; (iv) $\sin\left(\frac{\pi}{2} - \theta\right) = \cos\theta$; (v) $\sin(\theta + \phi) = \sin\theta\cos\phi + \cos\theta\sin\phi$; (vi) $\sin(\pi - \theta) = -\sin\theta$.

Proofs:

(i) $\cos(\theta + \phi) = \cos(\theta - (-\phi)) = \cos\theta\cos(-\phi) + \sin\theta\sin(-\phi)$ (by Problem 21)

$= \cos\theta\cos\phi - \sin\theta\sin\phi$ (by the Negative Identities). □

(ii) $\cos(\pi - \theta) = \cos\pi\cos\theta + \sin\pi\sin\theta = (-1)\cos\theta + 0 \cdot \sin\theta = -\cos\theta$. □

(iii) $\cos\left(\frac{\pi}{2} - \theta\right) = \cos\frac{\pi}{2}\cos\theta + \sin\frac{\pi}{2}\sin\theta = 0 \cdot \cos\theta + 1 \cdot \sin\theta = \sin\theta$. □

(iv) $\sin\left(\frac{\pi}{2} - \theta\right) = \cos\left(\frac{\pi}{2} - \left(\frac{\pi}{2} - \theta\right)\right) = \cos\left(\frac{\pi}{2} - \frac{\pi}{2} + \theta\right) = \cos\theta$. □

(v) $\sin(\theta + \phi) = \cos\left(\frac{\pi}{2} - (\theta + \phi)\right) = \cos\left(\left(\frac{\pi}{2} - \theta\right) - \phi\right)$

$= \cos\left(\frac{\pi}{2} - \theta\right)\cos\phi + \sin\left(\frac{\pi}{2} - \theta\right)\sin\phi = \sin\theta\cos\phi + \cos\theta\sin\phi$. □

(vi) $\sin(\pi - \theta) = \sin\pi\cos\theta + \cos\pi\sin\theta. = 0 \cdot \cos\theta + (-1)\sin\theta = -\sin\theta$. □

LEVEL 5

23. The Principle of Strong Induction is the following statement:

($\star\star$) Let $P(n)$ be a statement and suppose that (i) $P(0)$ is true and (ii) for all $k \in \mathbb{N}$, $\forall j \leq k\,(P(j)) \to P(k+1)$. Then $P(n)$ is true for all $n \in \mathbb{N}$.

Use the Principle of Mathematical Induction to prove the Principle of Strong Induction.

Proof: Let $P(n)$ be a statement such that $P(0)$ is true, and for all $k \in \mathbb{N}$, $\forall j \leq k\,(P(j)) \to P(k+1)$. Let $Q(n)$ be the statement $\forall j \leq n\,(P(j))$.

Base case: $Q(0) \equiv \forall j \leq 0(P(j)) \equiv P(0)$. Since $P(0)$ is true and $Q(0) \equiv P(0)$, $Q(0)$ is also true.

Inductive step: Suppose that $Q(k)$ is true. Then $\forall j \leq k\,(P(j))$ is true. Therefore, $P(k+1)$ is true. So $Q(k) \wedge P(k+1)$ is true. But notice that

$$Q(k+1) \equiv \forall j \leq k+1(P(j)) \equiv \forall j \leq k(P(j)) \wedge P(k+1) \equiv Q(k) \wedge P(k+1).$$

So, $Q(k+1)$ is true.

By the Principle of Mathematical Induction (($\star$) from Problem 11), $Q(n)$ is true for all $n \in \mathbb{N}$. This implies that $P(n)$ is true for all $n \in \mathbb{N}$. □

24. Use the Principle of Mathematical Induction to prove that for every $n \in \mathbb{N}$, if S is a set with $|S| = n$, then S has 2^n subsets. (Hint: Use Problem 16 from Problem Set 1.)

Proof: Base Case ($k = 0$): Let S be a set with $|S| = 0$. Then $S = \emptyset$, and the empty set has exactly 1 subset, namely itself. So, the number of subsets of S is $1 = 2^0$.

Inductive Step: Assume that for any set S with $|S| = k$, S has 2^k subsets.

Now, let A be a set with $|A| = k+1$, let d be any element from A, and let $S = A \setminus \{d\}$ (S is the set consisting of all elements of A except d). $|S| = k$, and so, by the inductive hypothesis, S has 2^k subsets. Let $B = \{X \mid X \subseteq A \wedge d \notin X\}$ and $C = \{X \mid X \subseteq A \wedge d \in X\}$. B is precisely the set of subsets of S, and so $|B| = 2^k$. By Problem 16 from Lesson 1, $|B| = |C|$ and therefore, $|C| = 2^k$. Also, B and C have no elements in common and every subset of A is in either B or C. So, the number of subsets of A is equal to $|B| + |C| = 2^k + 2^k = 2 \cdot 2^k = 2^1 \cdot 2^k = 2^{1+k} = 2^{k+1}$.

By the Principle of Mathematical Induction, given any $n \in \mathbb{N}$, if S is a set with $|S| = n$, then S has 2^n subsets. □

Notes: (1) Recall from Lesson 1 that $|S| = n$ means that the set S has n elements.

(2) Recall from Lesson 2 that if S is a set, then the **power set** of S is the set of subsets of S.

$$\mathcal{P}(S) = \{X \mid X \subseteq S\}$$

In this problem, we proved that a set with n elements has a power set with 2^n elements. Symbolically, we have

$$|S| = n \rightarrow |\mathcal{P}(S)| = 2^n.$$

25. Prove that addition of real numbers is well-defined and that the sum of two real numbers is a real number.

Proof: Suppose that $[(x_n)] = [(z_n)]$ and $[(y_n)] = [(w_n)]$. To prove that addition is well-defined, we need to show that $[(x_n + y_n)] = [(z_n + w_n)]$.

Let $k \in \mathbb{N}^+$. Since $[(x_n)] = [(z_n)]$, there is $K_1 \in \mathbb{N}$ such that $n > K_1$ implies $|x_n - z_n| < \frac{1}{2k}$. Since $[(y_n)] = [(w_n)]$, there is $K_2 \in \mathbb{N}$ such that $n > K_2$ implies $|y_n - w_n| < \frac{1}{2k}$. Let $K = \max\{K_1, K_2\}$. Let $n > K$. Since $K \geq K_1$, $n > K_1$, and therefore, we have $|x_n - z_n| < \frac{1}{2k}$. Since $K \geq K_2$, $n > K_2$, and therefore, we have $|y_n - w_n| < \frac{1}{2k}$. It follows that

$$|(x_n + y_n) - (z_n + w_n)| = |(x_n - z_n) + (y_n - w_n)| \leq |x_n - z_n| + |y_n - w_n| < \frac{1}{2k} + \frac{1}{2k} = \frac{1}{k}.$$

So, $[(x_n + y_n)] = [(z_n + w_n)]$, as desired.

We now prove that the sum of two real numbers is a real number. Let $[(x_n)]$ and $[(y_n)]$ be real numbers. Since the sum of two rational numbers is a rational number, for each $n \in \mathbb{N}$, we have $x_n + y_n \in \mathbb{Q}$. We need to show that $(x_n + y_n)$ is a Cauchy sequence. To see this, let $k \in \mathbb{N}^+$. Since (x_n) is a Cauchy sequence, there is $K_1 \in \mathbb{N}$ such that $m \geq n > K_1$ implies $|x_m - x_n| < \frac{1}{2k}$. Since (y_n) is a Cauchy sequence, there is $K_2 \in \mathbb{N}$ such that $m \geq n > K_2$ implies $|y_m - y_n| < \frac{1}{2k}$. Let $K = \max\{K_1, K_2\}$ and let $k \in \mathbb{N}^+$. Suppose that $m \geq n > K$. Since $K \geq K_1$, we have $|x_m - x_n| < \frac{1}{2k}$. Since $K \geq K_2$, we have $|y_m - y_n| < \frac{1}{2k}$. So,

$$|(x_m + y_m) - (x_n + y_n)| = |(x_m - x_n) + (y_m - y_n)| \leq |x_m - x_n| + |y_m - y_n| < \frac{1}{2k} + \frac{1}{2k} = \frac{1}{k}.$$

Therefore, $(x_n + y_n)$ is a Cauchy sequence. □

Problem Set 6

LEVEL 1

1. Show that there are exactly two monoids on the set $S = \{e, a\}$, where e is the identity. Which of these monoids are groups? Which of these monoids are commutative?

Solution: Let's let e be the identity. Since $e \star x = x \star e = x$ for all x in the monoid, we can easily fill out the first row and the first column of the table.

$\star$	e	a
e	e	a
a	a	$\boxdot$

Now, the entry labeled with $\boxdot$ must be either e or a because we need $\star$ to be a binary operation on S.

Case 1: If we let $\boxdot$ be a, we get the following table.

$\star$	e	a
e	e	a
a	a	a

Associativity holds because any computation of the form $(x \star y) \star z$ or $x \star (y \star z)$ will result in a if any of x, y, or z is a. So, all that is left to check is that $(e \star e) \star e = e \star (e \star e)$. But each side of that equation is equal to e.

So, with this multiplication table, $(S,\star)$ **is** a monoid.

This monoid is **not** a group because a has no inverse. Indeed, $a \star e = a \neq e$ and $a \star a = a \neq e$.

This monoid **is** commutative because $a \star e = a$ and $e \star a = a$.

Case 2: If we let $\boxdot$ be e, we get the following table.

$\star$	e	a
e	e	a
a	a	e

Let's check that associativity holds. There are eight instances to check.

$$(e \star e) \star e = e \star e = e \qquad e \star (e \star e) = e \star e = e$$
$$(e \star e) \star a = e \star a = a \qquad e \star (e \star a) = e \star a = a$$
$$(e \star a) \star e = a \star e = a \qquad e \star (a \star e) = e \star a = a$$
$$(a \star e) \star e = a \star e = a \qquad a \star (e \star e) = a \star e = a$$
$$(e \star a) \star a = a \star a = e \qquad e \star (a \star a) = e \star e = e$$
$$(a \star e) \star a = a \star a = e \qquad a \star (e \star a) = a \star a = e$$
$$(a \star a) \star e = e \star e = e \qquad a \star (a \star e) = a \star a = e$$
$$(a \star a) \star a = e \star a = a \qquad a \star (a \star a) = a \star e = a$$

So, with this multiplication table, $(S,\star)$ **is** a monoid.

Since $e \star e = e$, e is its own inverse. Since $a \star a = e$, a is also its own inverse. Therefore, each element of this monoid is invertible. It follows that this monoid **is** a group.

This monoid **is** commutative because $a \star e = a$ and $e \star a = a$.

2. The addition and multiplication tables below are defined on the set $S = \{0,1\}$. Show that $(S,+,\ \cdot)$ does **not** define a ring.

+	0	1
0	0	1
1	1	0

$\cdot$	0	1
0	1	0
1	0	1

Solution: We have $0(1+1) = 0 \cdot 0 = 1$ and $0 \cdot 1 + 0 \cdot 1 = 0 + 0 = 0$. So, $0(1+1) \neq 0 \cdot 1 + 0 \cdot 1$. Therefore, multiplication is **not** distributive over addition in S, and so, $(S,+,\ \cdot)$ does not define a ring.

Notes: (1) Both multiplication tables given are the same, except that we interchanged the roles of 0 and 1 (in technical terms, $(S,+)$ and $(S,\ \cdot)$ are **isomorphic**).

Both tables represent the unique table for a group with 2 elements. See Problem 1 above for details.

(2) Since $(S,+)$ is a commutative group and $(S,\ \cdot)$ is a monoid (in fact, it's a commutative group), we know that the only possible way $(S,+,\ \cdot)$ can fail to be a ring is for distributivity to fail.

3. The addition and multiplication tables below are defined on the set $S = \{0,1,2\}$. Show that $(S,+,\ \cdot)$ does **not** define a field.

+	0	1	2
0	0	1	2
1	1	2	0
2	2	0	1

$\cdot$	0	1	2
0	0	0	0
1	0	1	2
2	0	2	2

Solution: We have $2 \cdot 0 = 0$, $2 \cdot 1 = 2$, and $2 \cdot 2 = 2$. So, 2 has no multiplicative inverse, and therefore, $(S,+,\ \cdot)$ does **not** define a field.

Note: It's not difficult to check that $(S,+)$ is a group with identity 0 and $(S,\ \cdot)$ is a monoid with identity 1. However, $(S,+,\ \cdot)$ is not a ring, as distributivity fails. Here is a counterexample:

$$2(1+1) = 2 \cdot 2 = 2 \qquad\qquad 2 \cdot 1 + 2 \cdot 1 = 2 + 2 = 1$$

We could have used this computation to verify that $(S,+,\ \cdot)$ is not a field.

4. Let $F = \{0,1\}$, where $0 \neq 1$. Show that there is exactly one field $(F,+,\ \cdot)$, where 0 is the additive identity and 1 is the multiplicative identity.

Solution: Suppose that $(F,+,\ \cdot)$ is a field. Since $(F,+)$ is a commutative group, by Problem 1 above, the addition table must be the following.

+	0	1
0	0	1
1	1	0

Since $(F^*, \ \cdot)$ is a monoid and 1 is the multiplicative identity, we must have $1 \cdot 1 = 1$.

Now, if $0 \cdot 0 = 1$, then we have $1 = 0 \cdot 0 = 0(0+0) = 0 \cdot 0 + 0 \cdot 0 = 1 + 1 = 0$, a contradiction. So, $0 \cdot 0 = 0$.

If $0 \cdot 1 = 1$, then we have $1 = 0 \cdot 1 = (0+0) \cdot 1 = 0 \cdot 1 + 0 \cdot 1 = 1 + 1 = 0$, a contradiction. So, $0 \cdot 1 = 0$.

Finally, if $1 \cdot 0 = 1$, then we have $1 = 1 \cdot 0 = 1(0+0) = 1 \cdot 0 + 1 \cdot 0 = 1 + 1 = 0$, a contradiction. So, $1 \cdot 0 = 0$.

It follows that the addition and multiplication tables must be as follows:

+	0	1
0	0	1
1	1	0

·	0	1
0	0	0
1	0	1

Since we already know that $(F, +)$ is a commutative group and $(F^*, \ \cdot)$ is a monoid, all we need to verify is that distributivity and the multiplicative inverse property hold. Since $\cdot$ is commutative for S (by Problem 1 above), it suffices to verify left distributivity. We will do this by brute force. There are eight instances to check.

$$
\begin{array}{ll}
0(0+0) = 0 \cdot 0 = 0 & 0 \cdot 0 + 0 \cdot 0 = 0 + 0 = 0 \\
0(0+1) = 0 \cdot 1 = 0 & 0 \cdot 0 + 0 \cdot 1 = 0 + 0 = 0 \\
0(1+0) = 0 \cdot 1 = 0 & 0 \cdot 1 + 0 \cdot 0 = 0 + 0 = 0 \\
0(1+1) = 0 \cdot 0 = 0 & 0 \cdot 1 + 0 \cdot 1 = 0 + 0 = 0 \\
1(0+0) = 1 \cdot 0 = 0 & 1 \cdot 0 + 1 \cdot 0 = 0 + 0 = 0 \\
1(0+1) = 1 \cdot 1 = 1 & 1 \cdot 0 + 1 \cdot 1 = 0 + 1 = 1 \\
1(1+0) = 1 \cdot 1 = 1 & 1 \cdot 1 + 1 \cdot 0 = 1 + 0 = 1 \\
1(1+1) = 1 \cdot 0 = 0 & 1 \cdot 1 + 1 \cdot 1 = 1 + 1 = 0
\end{array}
$$

So, we see that left distributivity holds, and therefore $(S, +, \ \cdot)$ is a commutative ring.

Since $1 \cdot 1 = 1$, the multiplicative inverse property holds, and it follows that $(F, +, \ \cdot)$ is a field.

LEVEL 2

5. Let $G = \{e, a, b\}$ and let $(G, \star)$ be a group with identity element e. Draw a multiplication table for $(G, \star)$.

Solution: Since $e \star x = x \star e = x$ for all x in the group, we can easily fill out the first row and the first column of the table.

$\star$	e	a	b
e	e	a	b
a	a	⊡	
b	b		

Now, the entry labeled with ⊡ must be either e or b because a is already in that row. If it were e, then the final entry in the row would be b giving two b's in the last column. Therefore, the entry labeled with ⊡ must be b.

$\star$	e	a	b
e	e	a	b
a	a	b	
b	b		

Since the same element cannot be repeated in any row or column, the rest of the table is now determined.

$\star$	$\boldsymbol{e}$	$\boldsymbol{a}$	$\boldsymbol{b}$
$\boldsymbol{e}$	$\boldsymbol{e}$	$\boldsymbol{a}$	$\boldsymbol{b}$
$\boldsymbol{a}$	$\boldsymbol{a}$	$\boldsymbol{b}$	$\boldsymbol{e}$
$\boldsymbol{b}$	$\boldsymbol{b}$	$\boldsymbol{e}$	$\boldsymbol{a}$

Notes: (1) Why can't the same element appear twice in any row? Well if x appeared twice in the row corresponding to y, that would mean that there are elements z and w with $\boldsymbol{z \neq w}$ such that $y \star z = x$ and $y \star w = x$. So, $y \star z = y \star w$. We can multiply each side of the equation on the left by y^{-1} (the inverse of y) to get $y^{-1} \star (y \star z) = y^{-1} \star (y \star w)$. By associativity, $(y^{-1} \star y) \star z = (y^{-1} \star y) \star w$. Now, $y^{-1} \star y = e$ by the inverse property. So, we have $e \star z = e \star w$. Finally, since e is an identity, $\boldsymbol{z = w}$. This contradiction establishes that no element x can appear twice in the same row of a group multiplication table.

A similar argument can be used to show that the same element cannot appear twice in any column.

(2) The argument given in Note 1 used all the group properties (associativity, identity, and inverse). What if we remove one of the properties. For example, what about the multiplication table for a monoid? Can an element appear twice in a row or column? I leave this as an optional exercise.

(3) In Note 1 above, we showed that in the multiplication table for a group, the same element cannot appear as the output more than once in any row or column. We can also show that every element must appear in every row and column. Let's show that the element y must appear in the row corresponding to x. We are looking for an element z such that $x \star z = y$. Well, $z = x^{-1} \star y$ works. Indeed, we have $x \star (x^{-1} \star y) = (x \star x^{-1}) \star y = e \star y = y$.

(4) Using Notes 1 and 3, we see that each element of a group appears exactly once in every row and column of the group's multiplication table.

(5) We have shown that there is essentially just one group of size 3, namely the one given by the table that we produced. Any other group with 3 elements will look exactly like this one, except for possibly the names of the elements. In technical terms, we say that any two groups of order 3 are **isomorphic**.

(6) Observe that in the table we produced, $b = a \star a$. We will generally abbreviate $a \star a$ as a^2. So, another way to draw the table is as follows:

$\star$	e	a	a^2
e	e	a	a^2
a	a	a^2	e
a^2	a^2	e	a

This group is the **cyclic group of order 3**. We call it **cyclic** because the group consists of all powers of the single element a (the elements are a, a^2, and $a^3 = a^0 = e$). The **order** is the number of elements in the group.

6. Prove that in any monoid $(M,\star)$, the identity element is unique.

Proof: Let $(M,\star)$ be a monoid and suppose that e and f are both identity elements in M. Then, we have $f = e \star f = e$. Since we have shown f and e to be equal, there is only one identity element. □

Notes: (1) The word "unique" means that there is only one. In mathematics, we often show that an object is unique by starting with two such objects and then arguing that they must actually be the same. Notice that in the proof above, when we said that e and f are both identity elements, we never insisted that they be *distinct* identity elements. And in fact, the end of the argument shows that they are not distinct.

(2) $e \star f = f$ because e is an identity element and $e \star f = e$ because f is an identity element.

7. Let $(F, +, \cdot)$ be a field. Prove each of the following:
   - (i) If $a, b \in F$ with $a + b = b$, then $a = 0$.
   - (ii) If $a \in F$, $b \in F^*$, and $ab = b$, then $a = 1$.
   - (iii) If $a \in F$, then $a \cdot 0 = 0$.
   - (iv) If $a \in F^*$, $b \in F$, and $ab = 1$, then $b = \frac{1}{a}$.
   - (v) If $a, b \in F$ and $ab = 0$, then $a = 0$ or $b = 0$.
   - (vi) If $a \in F$, then $-a = -1a$.
   - (vii) $(-1)(-1) = 1$.

Proofs:

(i) Let $a, b \in F$ with $a + b = b$. Then we have

$$a = a + 0 = a + \big(b + (-b)\big) = (a + b) + (-b) = b + (-b) = 0.$$ □

(ii) Let $a \in F$, $b \in F^*$, and $ab = b$. Then we have

$$a = a \cdot 1 = a(bb^{-1}) = (ab)b^{-1} = bb^{-1} = 1.$$ □

(iii) Let $a \in F$. Then $a \cdot 0 + a = a \cdot 0 + a \cdot 1 = a(0 + 1) = a \cdot 1 = a$. By (i), $a \cdot 0 = 0$. □

(iv) Let $a \in F^*$, $b \in F$, and $ab = 1$. Then $b = 1b = (a^{-1}a)b = a^{-1}(ab) = a^{-1} \cdot 1 = a^{-1} = \frac{1}{a}$. □

(v) Let $a, b \in F$ and $ab = 0$. Assume that $a \neq 0$. Then $b = 1b = (a^{-1}a)b = a^{-1}(ab) = a^{-1} \cdot 0$. By (iii), $a^{-1} \cdot 0 = 0$. So, $b = 0$. □

(vi) Let $a \in F$. Then $-1a + a = a(-1) + a \cdot 1 = a(-1+1) = a \cdot 0 = 0$ (by (iii)). So, $-1a$ is the additive inverse of a. Thus, $-1a = -a$. □

(vii) $(-1)(-1) + (-1) = (-1)(-1) + (-1) \cdot 1 = (-1)(-1+1) = (-1)(0) = 0$ (by (iii)). So, we see that $(-1)(-1)$ is the additive inverse of -1. Therefore, $(-1)(-1) = -(-1)$. □

8. Let $(F, +, \cdot)$ be a field with $\mathbb{N} \subseteq F$. Prove that $\mathbb{Q} \subseteq F$.

Proof: Let $n \in \mathbb{Z}$. If $n \in \mathbb{N}$, then $n \in F$ because $\mathbb{N} \subseteq F$. If $n \notin \mathbb{N}$, then $-n \in \mathbb{N}$. So, $-n \in F$. Since F is a field, we have $n = -(-n) \in F$. For each $n \in \mathbb{Z}^*$, $\frac{1}{n} = n^{-1} \in F$ because $n \in F$ and the multiplicative inverse property holds in F. Now, let $\frac{m}{n} \in \mathbb{Q}$. Then $m \in \mathbb{Z}$ and $n \in \mathbb{Z}^*$. Since $\mathbb{Z} \subseteq F$, $m \in F$. Since $n \in \mathbb{Z}^*$, we have $\frac{1}{n} \in F$. Therefore, $\frac{m}{n} = \frac{m \cdot 1}{1 \cdot n} = \frac{m}{1} \cdot \frac{1}{n} = m\left(\frac{1}{n}\right) \in F$ because F is closed under multiplication. Since $\frac{m}{n}$ was an arbitrary element of $\mathbb{Q}$, we see that $\mathbb{Q} \subseteq F$. □

LEVEL 3

9. Assume that a group $(G, \star)$ of order 4 exists with $G = \{e, a, b, c\}$, where e is the identity, $a^2 = b$ and $b^2 = e$. Construct the table for the operation of such a group.

Solution: Since $e \star x = x \star e = x$ for all x in the group, we can easily fill out the first row and the first column of the table.

$\star$	e	a	b	c
e	e	a	b	c
a	a			
b	b			
c	c			

We now add in $a \star a = a^2 = b$ and $b \star b = b^2 = e$.

$\star$	e	a	b	c
e	e	a	b	c
a	a	b	⊡	
b	b		e	
c	c			

Now, the entry labeled with ⊡ cannot be a or b because a and b appear in that row. It also cannot be e because e appears in that column. Therefore, the entry labeled with ⊡ must be c. It follows that the entry to the right of ⊡ must be e, and the entry at the bottom of the column must be a.

$\star$	e	a	b	c
e	e	a	b	c
a	a	b	c	e
b	b	⊙	e	
c	c		a	

Now, the entry labeled with ⊙ cannot be b or e because b and e appear in that row. It also cannot be a because a appears in that column. Therefore, the entry labeled with ⊙ must be c. The rest of the table is then determined.

$\star$	e	a	b	c
e	e	a	b	c
a	a	b	c	e
b	b	c	e	a
c	c	e	a	b

Note: Observe that in the table we produced, $b = a \star a = a^2$ and $c = b \star a = a^2 \star a = a^3$. So, another way to draw the table is as follows:

$\star$	e	a	a^2	a^3
e	e	a	a^2	a^3
a	a	a^2	a^3	e
a^2	a^2	a^3	e	a
a^3	a^3	e	a	a^2

This group is the **cyclic group of order 4**.

10. Prove that in any group $(G,\star)$, each element has a unique inverse.

Proof: Let $a \in G$ and suppose that $b, c \in G$ are both inverses of a. We will show that b and c must be the same. We have $c = c \star e = c \star (a \star b) = (c \star a) \star b = e \star b = b$. □

Notes: (1) $c = c \star e$ because e is an identity element.

(2) $e = a \star b$ because b is an inverse of a. So, $c \star e = c \star (a \star b)$.

(3) $c \star (a \star b) = (c \star a) \star b$ by associativity of $\star$.

(4) $c \star a = e$ because c is an inverse of a. So, $(c \star a) \star b = e \star b$.

(5) $e \star b = b$ because e is an identity element.

11. Let $(F, +, \cdot, \leq)$ be an ordered field. Prove each of the following:

(i) If $a, b \in F^+$ and $a > b$, then $\frac{1}{a} < \frac{1}{b}$.

(ii) If $a, b \in F$, then $a \geq b$ if and only if $-a \leq -b$.

Proofs:

(i) Let $a, b \in F^+$ and $a > b$. Then $a - b = a + (-b) > b + (-b) = 0$ (we used Order Property 1 here). So, $\frac{1}{b} - \frac{1}{a} = \frac{1}{ab}(a - b)$. Since $a, b \in F^+$, $ab \in F^+$ by Order Property (2). So, $\frac{1}{ab} \in F^+$ by Theorem 6.8. Since $\frac{1}{ab} > 0$ and $a - b > 0$, we have $\frac{1}{b} - \frac{1}{a} = \frac{1}{ab}(a - b) > 0$. So, $\frac{1}{b} > \frac{1}{a}$, or equivalently, $\frac{1}{a} < \frac{1}{b}$. □

(ii) Let $a, b \in F$. Then $a \geq b$ if and only if $a - b \geq 0$ if and only if $-(a-b) \leq 0$ if and only if $-1(a + (-b)) \leq 0$ if and only if $-1a - 1(-b) \leq 0$ if and only if $-a - (-b) \leq 0$ if and only if $-a < -b$. □

12. Let $(F, +, \cdot)$ be a field. Show that $(F, \cdot)$ is a commutative monoid.

Proof: Let $(F, +, \cdot)$ be a field. Then $\cdot$ is a binary operation on F and $(F^*, \cdot)$ is a commutative group.

We first show that if $a \in F$, then $0a = 0$. To see this, observe that

$$0a + a = 0a + 1a = (0 + 1)a = 1a = a.$$

By Problem 7, part (i), $0a = 0$.

Let $x, y \in F$. If $x, y \in F^*$, then $xy = yx$. If $x = 0$, then $xy = 0y = 0$ by the previous result, and $yx = y \cdot 0 = 0$ by Problem 7, part (iii) above. If $y = 0$, then $xy = x \cdot 0 = 0$ by Problem 7, part (iii) above, and $yx = 0x = 0$ by the previous result. In all cases, we have $xy = yx$.

Next, let $x, y, z \in F$. If $x, y, z \in F^*$, then $(xy)z = x(yz)$. If $x = 0$, then by the previous result, we have $(xy)z = (0y)z = 0z = 0$ and $x(yz) = 0(yz) = 0$. If $y = 0$, by Problem 7, part (iii) and the previous result, we have $(xy)z = (x \cdot 0)z = 0z = 0$ and $x(yz) = x(0z) = x \cdot 0 = 0$. If $z = 0$, we have $(xy)z = (xy) \cdot 0 = 0$ and $x(yz) = x(y \cdot 0) = x \cdot 0 = 0$. In all cases, we have $(xy)z = x(yz)$.

Let $x \in F$. If $x \in F^*$, then $1x = x \cdot 1 = x$. If $x = 0$, then by Problem 7, part (iii), $1x = 1 \cdot 0 = 0$ and by the previous result, $x \cdot 1 = 0 \cdot 1 = 0$. In all cases, we have $1x = x \cdot 1 = x$.

Therefore, $(F, \cdot)$ is a commutative monoid. □

LEVEL 4

13. Let $(G, \star)$ be a group with $a, b \in G$, and let a^{-1} and b^{-1} be the inverses of a and b, respectively. Prove

(i) $(a \star b)^{-1} = b^{-1} \star a^{-1}$.

(ii) the inverse of a^{-1} is a.

Proof of (i): Let $a, b \in G$. Then we have

$$(a \star b) \star (b^{-1} \star a^{-1}) = a \star (b \star (b^{-1} \star a^{-1})) = a \star ((b \star b^{-1}) \star a^{-1}) = a \star (e \star a^{-1}) = a \star a^{-1} = e$$

and

$$(b^{-1} \star a^{-1}) \star (a \star b) = b^{-1} \star (a^{-1} \star (a \star b)) = b^{-1} \star ((a^{-1} \star a) \star b) = b^{-1} \star (e \star b) = b^{-1} \star b = e.$$

So, $(a \star b)^{-1} = (b^{-1} \star a^{-1})$. □

Notes: (1) For the first and second equalities we used the associativity of $\star$ in G.

(2) For the third equality, we used the inverse property of $\star$ in G.

(3) For the fourth equality, we used the identity property of $\star$ in G.

(4) For the last equality, we again used the inverse property of $\star$ in G.

(5) Since multiplying $a \star b$ on either side by $b^{-1} \star a^{-1}$ results in the identity element e, it follows that $b^{-1} \star a^{-1}$ is the inverse of $a \star b$.

(6) In a group, to verify that an element h is the inverse of an element g, it suffices to show that $g \star h = e$ **or** $h \star g = e$. In other words, we can prove that $g \star h = e \to h \star g = e$ and we can prove that $h \star g = e \to g \star h = e$.

For a proof that $g \star h = e \to h \star g = e$, suppose that $g \star h = e$ and k is the inverse of g. Then $g \star k = k \star g = e$. Since $g \star h = e$ and $g \star k = e$, we have $g \star h = g \star k$. By multiplying by g^{-1} on each side of this equation, and using associativity, the inverse property, and the identity property, we get $h = k$. So, h is in fact the inverse of g.

Proving that $h \star g = e \to g \star h = e$ is similar. Thus, in the solution above, we need only show one of the sequences of equalities given. The second one follows for free.

Proof of (ii): Let $a \in G$. Since a^{-1} is the inverse of a, we have $a \star a^{-1} = a^{-1} \star a = e$. But this sequence of equations also says that a is the inverse of a^{-1}. □

14. Prove that there is no smallest positive real number.

Proof: Let $x \in \mathbb{R}^+$ and let $y = \frac{1}{2}x$. By Theorem 6.8, $\frac{1}{2} > 0$. So, by Order Property (2), $y > 0$.

Now, $x - y = x - \frac{1}{2}x = 1x - \frac{1}{2}x = \left(1 - \frac{1}{2}\right)x = \left(\frac{2}{2} - \frac{1}{2}\right)x = \frac{1}{2}x > 0$. So, $x > y$. It follows that y is a positive real number smaller than x. Since x was an arbitrary positive real number, there is no smallest positive real number. □

15. Let a be a nonnegative real number. Prove that $a = 0$ if and only if a is less than every positive real number. (Note: a nonnegative means $a \geq 0$.)

Proof: Let a be a nonnegative real number.

First suppose that $a = 0$. Let ϵ be a positive real number, so that $\epsilon > 0$. Then by direct substitution, $\epsilon > a$, or equivalently $a < \epsilon$. Since ϵ was an arbitrary positive real number, we have shown that a is less than every positive real number.

Now, suppose that a is less than every positive real number. Assume towards contradiction that $a \neq 0$. Then $a > 0$ (because a is nonnegative). Let $\epsilon = \frac{1}{2}a$. By the same reasoning used in Problem 14 above, we have that ϵ is a positive real number with $a > \epsilon$. This contradicts our assumption that a is less than every positive real number. □

16. Prove that every rational number can be written in the form $\frac{m}{n}$, where $m \in \mathbb{Z}$, $n \in \mathbb{Z}^*$, and at least one of m or n is **not** even.

Proof: Let x be a rational number. Then there are $a \in \mathbb{Z}$ and $b \in \mathbb{Z}^*$ such that $x = \frac{a}{b}$. Let j be the largest integer such that 2^j divides a and let k be the largest integer such that 2^k divides b. Since, 2^j divides a, there is $c \in \mathbb{Z}$ such that $a = 2^j c$. Since, 2^k divides b, there is $d \in \mathbb{Z}$ such that $b = 2^k d$.

Observe that c is odd. Indeed, if c were even, then there would be an integer s such that $c = 2s$. But then $a = 2^j c = 2^j(2s) = (2^j \cdot 2)s = (2^j \cdot 2^1)s = 2^{j+1}s$. So, 2^{j+1} divides a, contradicting the maximality of j.

Similarly, d is odd.

So, we have $x = \frac{a}{b} = \frac{2^j c}{2^k d}$.

If $j \geq k$, then, $j - k \geq 0$ and $x = \frac{2^j c}{2^k d} = \frac{2^{j-k} c}{d}$. Let $m = 2^{j-k} c$ and $n = d$. Then $x = \frac{m}{n}$, $m \in \mathbb{Z}$ (because $\mathbb{Z}$ is closed under multiplication), $n \in \mathbb{Z}^*$ (if $n = 0$, then $b = 2^k d = 2^k n = 2^k \cdot 0 = 0$), and $n = d$ is odd.

If $j < k$, then $k - j > 0$ and $x = \frac{2^j c}{2^k d} = \frac{c}{2^{k-j} d}$. Let $m = c$ and $n = 2^{k-j} d$. Then $x = \frac{m}{n}$, $m = c \in \mathbb{Z}$, $n \in \mathbb{Z}^*$ (because $\mathbb{Z}$ is closed under multiplication, and if n were 0, then d would be 0, and then b would be 0), and $m = c$ is odd. □

LEVEL 5

17. Prove that $(\mathbb{Q}, +, \cdot, \leq)$ and $(\mathbb{R}, +, \cdot, \leq)$ are ordered fields. Also prove that $(\mathbb{C}, +, \cdot)$ is field that cannot be ordered.

Proof: We first prove that $(\mathbb{Q}, +)$ is a commutative group.

(Closure) Let $x, y \in \mathbb{Q}$. Then there exist $a, c \in \mathbb{Z}$ and $b, d \in \mathbb{Z}^*$ such that $x = \frac{a}{b}$ and $y = \frac{c}{d}$. We have $x + y = \frac{a}{b} + \frac{c}{d} = \frac{ad+bc}{bd}$. Since $\mathbb{Z}$ is closed under multiplication, $ad \in \mathbb{Z}$ and $bc \in \mathbb{Z}$. Since $\mathbb{Z}$ is closed under addition, $ad + bc \in \mathbb{Z}$. Since $\mathbb{Z}^*$ is closed under multiplication, $bd \in \mathbb{Z}^*$. Therefore, $x + y \in \mathbb{Q}$.

(Associativity) Let $x, y, z \in \mathbb{Q}$. Then there exist $a, c, e \in \mathbb{Z}$ and $b, d, f \in \mathbb{Z}^*$ such that $x = \frac{a}{b}$, $y = \frac{c}{d}$, and $z = \frac{e}{f}$. Since multiplication and addition are associative in $\mathbb{Z}$, multiplication is (both left and right) distributive over addition in $\mathbb{Z}$ (see the Note below), and multiplication is associative in $\mathbb{Z}^*$, we have

$$(x + y) + z = \left(\frac{a}{b} + \frac{c}{d}\right) + \frac{e}{f} = \frac{ad + bc}{bd} + \frac{e}{f} = \frac{(ad + bc)f + (bd)e}{(bd)f} = \frac{\big((ad)f + (bc)f\big) + (bd)e}{(bd)f}$$

$$= \frac{a(df) + \big(b(cf) + b(de)\big)}{b(df)} = \frac{a(df) + b(cf + de)}{b(df)} = \frac{a}{b} + \frac{cf + de}{df} = \frac{a}{b} + \left(\frac{c}{d} + \frac{e}{f}\right) = x + (y + z).$$

(Identity) Let $\overline{0} = \frac{0}{1}$. We show that $\overline{0}$ is an identity for $(\mathbb{Q}, +)$. Let $x \in \mathbb{Q}$. Then there exist $a \in \mathbb{Z}$ and $b \in \mathbb{Z}^*$ such that $x = \frac{a}{b}$. Since 0 is an identity for $\mathbb{Z}$, and $0 \cdot x = x \cdot 0 = 0$ for all $x \in \mathbb{Z}$, we have

$$x+\overline{0}=\frac{a}{b}+\frac{0}{1}=\frac{a\cdot 1+b\cdot 0}{b\cdot 1}=\frac{a+0}{b}=\frac{a}{b}=x \text{ and } \overline{0}+x=\frac{0}{1}+\frac{a}{b}=\frac{0b+1a}{1b}=\frac{0+a}{b}=\frac{a}{b}=x.$$

(Inverse) Let $x \in \mathbb{Q}$. Then there exist $a \in \mathbb{Z}$ and $b \in \mathbb{Z}^*$ such that $x=\frac{a}{b}$. Let $y=\frac{-1a}{b}$. Since $\mathbb{Z}$ is closed under multiplication, $-1a \in \mathbb{Z}$. So, $y \in \mathbb{Q}$. Since multiplication is associative and commutative in $\mathbb{Z}$ and $(-1)n=-n$ for all $n \in \mathbb{Z}$, we have

$$x+y=\frac{a}{b}+\frac{-1a}{b}=\frac{ab+b(-1a)}{b\cdot b}=\frac{ab+(-1a)b}{b^2}=\frac{ab+(-1)(ab)}{b^2}=\frac{ab-ab}{b^2}=\frac{0}{b^2}=\overline{0}$$

$$y+x=\frac{-1a}{b}+\frac{a}{b}=\frac{(-1a)b+ba}{b\cdot b}=\frac{-1(ab)+ab}{b^2}=\frac{-ab+ab}{b^2}=\frac{0}{b^2}=\overline{0}$$

So, y is the additive inverse of x.

(Commutativity) Let $x, y \in \mathbb{Q}$. Then there exist $a, c \in \mathbb{Z}$ and $b, d \in \mathbb{Z}^*$ such that $x=\frac{a}{b}$ and $y=\frac{c}{d}$. Since multiplication and addition are commutative in $\mathbb{Z}$, and multiplication is commutative in $\mathbb{Z}^*$, we have

$$x+y=\frac{a}{b}+\frac{c}{d}=\frac{ad+bc}{bd}=\frac{bc+ad}{db}=\frac{cb+da}{db}=\frac{c}{d}+\frac{a}{b}=y+x.$$

So, $(\mathbb{Q},+)$ is a commutative group.

We next prove that $(\mathbb{Q}\setminus\{0\},\ \cdot)$ is a commutative group.

(Closure) Let $x, y \in \mathbb{Q}^*$. Then there exist $a, b, c, d \in \mathbb{Z}^*$ such that $x=\frac{a}{b}$ and $y=\frac{c}{d}$. We have $xy=\frac{a}{b}\cdot\frac{c}{d}=\frac{ac}{bd}$. Since $\mathbb{Z}^*$ is closed under multiplication, $ac, bd \in \mathbb{Z}^*$. Therefore, $xy \in \mathbb{Q}^*$.

(Associativity) Let $x, y, z \in \mathbb{Q}^*$. Then there exist $a, b, c, d, e, f \in \mathbb{Z}^*$ such that $x=\frac{a}{b}$, $y=\frac{c}{d}$, and $z=\frac{e}{f}$. Since multiplication is associative in $\mathbb{Z}^*$, we have

$$(xy)z=\left(\frac{a}{b}\cdot\frac{c}{d}\right)\frac{e}{f}=\left(\frac{ac}{bd}\right)\frac{e}{f}=\frac{(ac)e}{(bd)f}=\frac{a(ce)}{b(df)}=\frac{a}{b}\left(\frac{ce}{df}\right)=\frac{a}{b}\left(\frac{c}{d}\cdot\frac{e}{f}\right)=x(yz).$$

(Identity) Let $\overline{1}=\frac{1}{1}$. We show that $\overline{1}$ is an identity for $(\mathbb{Q}^*,\ \cdot)$. Let $x \in \mathbb{Q}^*$. Then there exist $a, b \in \mathbb{Z}^*$ such that $x=\frac{a}{b}$. Since 1 is an identity for $\mathbb{Z}^*$, we have

$$x\cdot\overline{1}=\frac{a}{b}\cdot\frac{1}{1}=\frac{a\cdot 1}{b\cdot 1}=\frac{a}{b}=x \text{ and } \overline{1}x=\frac{1}{1}\cdot\frac{a}{b}=\frac{1a}{1b}=\frac{a}{b}=x.$$

(Inverse) Let $x \in \mathbb{Q}^*$. Then there exist $a, b \in \mathbb{Z}^*$ such that $x=\frac{a}{b}$. Let $y=\frac{b}{a}$. Then $y \in \mathbb{Q}^*$ (note that $a \neq 0$). Since multiplication is commutative in $\mathbb{Z}^*$, we have

$$xy=\frac{a}{b}\cdot\frac{b}{a}=\frac{ab}{ba}=\frac{ab}{ab}=\frac{1}{1}=\overline{1}.$$

So, y is the multiplicative inverse of x.

(Commutativity) Let $x,\ y \in \mathbb{Q}^*$. Then there exist $a, b, c, d \in \mathbb{Z}^*$ such that $x = \frac{a}{b}$ and $y = \frac{c}{d}$. Since multiplication is commutative in $\mathbb{Z}^*$, we have

$$xy = \frac{a}{b} \cdot \frac{c}{d} = \frac{ac}{bd} = \frac{ca}{db} = \frac{c}{d} \cdot \frac{a}{b} = yx.$$

So, $(\mathbb{Q}^*,\ \cdot)$ is a commutative group.

Now we prove that multiplication is distributive over addition in $\mathbb{Q}$.

(Distributivity) Let $x, y, z \in \mathbb{Q}$. Then there exist $a, c, e \in \mathbb{Z}$ and $b, d, f \in \mathbb{Z}^*$ such that $x = \frac{a}{b}$, $y = \frac{c}{d}$, and $z = \frac{e}{f}$. Let's start with left distributivity.

$$x(y+z) = \frac{a}{b}\left(\frac{c}{d} + \frac{e}{f}\right) = \frac{a}{b}\left(\frac{cf+de}{df}\right) = \frac{a(cf+de)}{b(df)}$$

$$xy + xz = \frac{a}{b} \cdot \frac{c}{d} + \frac{a}{b} \cdot \frac{e}{f} = \frac{ac}{bd} + \frac{ae}{bf} = \frac{(ac)(bf) + (bd)(ae)}{(bd)(bf)}$$

We need to verify that $\frac{(ac)(bf)+(bd)(ae)}{(bd)(bf)} = \frac{a(cf+de)}{b(df)}$.

Since $\mathbb{Z}$ is a ring, $(ac)(bf) + (bd)(ae) = bacf + bade = ba(cf + de)$ (see Note 1 below).

Since multiplication is associative and commutative in $\mathbb{Z}^*$, we have

$$(bd)(bf) = b\big(d(bf)\big) = b\big((db)f\big) = b\big((bd)f\big) = b\big(b(df)\big).$$

So, $\frac{(ac)(bf)+(bd)(ae)}{(bd)(bf)} = \frac{ba(cf+de)}{b\left(b(df)\right)} = \frac{a(cf+de)}{b(df)}$.

For right distributivity, we can use left distributivity together with the commutativity of multiplication in $\mathbb{Q}$.

$$(y+z)x = x(y+z) = xy + xz = yx + zx \qquad \square$$

Notes: (1) We skipped many steps when verifying $(ac)(bf) + (bd)(ae) = ba(cf + de)$. The dedicated reader may want to verify this equality carefully, making sure to use only the fact that $\mathbb{Z}$ is a ring, and making a note of which ring property is being used at each step.

(2) In the very last step of the proof, we cancelled one b in the numerator of the fraction with b in the denominator of the fraction. In general, if $j \in \mathbb{Z}$ and $m, k \in \mathbb{Z}^*$, then $\frac{mj}{mk} = \frac{j}{k}$. To verify that this is true, simply observe that since $\mathbb{Z}$ is a ring, we have $(mj)k = m(jk) = m(kj) = (mk)j$.

We next prove the order properties. Let $\frac{a}{b}, \frac{c}{d}, \frac{e}{f} \in \mathbb{Q}$ and assume that $\frac{a}{b} \leq \frac{c}{d}$. Then $ad \leq bc$. Since $\mathbb{Z}$ is an ordered ring, we get $(af + be)(df) = afdf + bedf = adff + bedf \leq bcff + bedf = bf(cf + de)$. Therefore, $\frac{a}{b} + \frac{e}{f} = \frac{af+be}{bf} \leq \frac{cf+de}{df} = \frac{c}{d} + \frac{e}{f}$.

Next, let $\frac{a}{b}, \frac{c}{d} \in \mathbb{Q}$ with $0 \leq \frac{a}{b}$ and $0 \leq \frac{c}{d}$. We may assume that $a, b, c, d \geq 0$. Then $\frac{a}{b} \cdot \frac{c}{d} = \frac{ac}{bd} \geq 0$.

It follows that $(\mathbb{Q}, +, \cdot, \leq)$ is an ordered field.

We proved that $(\mathbb{R}, +)$ is a commutative group in part 6 of Example 6.3. We next prove that $(\mathbb{R} \setminus \{0\}, \cdot)$ is a commutative group.

(Closure) This is Problem 25 from Problem Set 5.

(Associativity) To see that $\cdot$ is associative in $\mathbb{R}^*$, we use the associativity of $\cdot$ in $\mathbb{Q}$. If $[(x_n)], [(y_n)], [(z_n)] \in \mathbb{R}^*$, then

$$([(x_n)] \cdot [(y_n)]) \cdot [(z_n)] = [(x_n \cdot y_n)] \cdot [(z_n)] = \left[\left((x_n \cdot y_n) \cdot z_n\right)\right]$$
$$= \left[\left(x_n \cdot (y_n \cdot z_n)\right)\right] = [x_n] \cdot [(y_n \cdot z_n)] = [(x_n)] \cdot ([(y_n)] \cdot [(z_n)]).$$

(Identity) To see that $[(1)]$ is the multiplicative identity, using the fact that 1 is the additive identity in $\mathbb{Q}$, we have for $[(x_n)] \in \mathbb{R}$,

$$[(1)] \cdot [(x_n)] = [(1 \cdot x_n)] = [(x_n)] \text{ and } [(x_n)] \cdot [(1)] = [(x_n \cdot 1)] = [(x_n)].$$

(Inverse) The inverse of the real number $[(x_n)]$ is $[(y_n)]$, where for each $n \in \mathbb{N}$, $y_n = \frac{1}{x_n}$ if $x_n \neq 0$ and $y_n = 0$ if $x_n = 0$. We have that $[(x_n)] \cdot [(y_n)] = [(z_n)]$ and $[(y_n)] + [(x_n)] = [(z_n)]$, where z_n is 0 or 1 for all $n \in \mathbb{N}$. We claim that $[(z_n)] = [(1)]$. To see this, note that since $[(x_n)] \neq [(0)]$, there is a $K > 0$ such that for $n > N$, $x_n \neq 0$.

(Commutativity) To see that $\cdot$ is commutative in $\mathbb{R}^*$, we use the commutativity of $\cdot$ in $\mathbb{Q}$. If $[(x_n)], [(y_n)] \in \mathbb{R}^*$, then $[(x_n)] \cdot [(y_n)] = [(x_n y_n)] = [(y_n x_n)] = [(y_n)] \cdot [(x_n)]$.

(Distributivity) Distributivity is similar to commutativity and associativity.

We next prove the order properties. Let $[(x_n)], [(y_n)], [(z_n)] \in \mathbb{R}$ and assume that $[(x_n)] \leq [(y_n)]$. Then there is $K \in \mathbb{N}$ such that $n > K$ implies $x_n \leq y_n$. Since $\mathbb{Q}$ is an ordered ring, it follows that $n > K$ implies $x_n + z_n \leq y_n + z_n$. So, $[(x_n)] + [(z_n)] = [(x_n + z_n)] \leq [(y_n + z_n)] = [(y_n)] + [(z_n)]$.

Next, let $[(x_n)], [(y_n)] \in \mathbb{R}$ with $[(0)] \leq [(x_n)]$ and $[(0)] \leq [(y_n)]$. Then there is $K_1 \in \mathbb{N}$ such that $n > K_1$ implies $x_n \geq 0$ and there is $K_2 \in \mathbb{N}$ such that $n > K_2$ implies $y_n \geq 0$. Let $K = \max\{K_1, K_2\}$ and let $n > K$. Since $K \geq K_1$, $x_n \geq 0$. Since $K \geq K_2$, $y_n \geq 0$. Since $\mathbb{Q}$ is an ordered ring, $x_n \cdot y_n \geq 0$. Therefore, we have $[(x_n)] \cdot [(y_n)] \geq [(0)]$.

It follows that $(\mathbb{R}, +, \cdot, \leq)$ is an ordered field.

We now prove that $(\mathbb{C}, +)$ is a commutative group.

(Closure) Let $z, w \in \mathbb{C}$. Then there are $a, b, c, d \in \mathbb{R}$ such that $z = a + bi$ and $w = c + di$. By definition, $z + w = (a + bi) + (c + di) = (a + c) + (b + d)i$. Since $\mathbb{R}$ is closed under addition, $a + b \in \mathbb{R}$ and $c + d \in \mathbb{R}$. Therefore, $z + w \in \mathbb{C}$.

(Associativity) Let $z, w, v \in \mathbb{C}$. Then there are $a, b, c, d, e, f \in \mathbb{R}$ such that $z = a + bi$, $w = c + di$, and $v = e + fi$. Since addition is associative in $\mathbb{R}$, we have

$$(z + w) + v = \left((a + bi) + (c + di)\right) + (e + fi) = \left((a + c) + (b + d)i\right) + (e + fi)$$
$$= \left((a + c) + e\right) + \left((b + d) + f\right)i = \left(a + (c + e)\right) + \left(b + (d + f)\right)i$$
$$= (a + bi) + \left((c + e) + (d + f)i\right) = (a + bi) + \left((c + di) + (e + fi)\right) = z + (w + v).$$

(Commutativity) Let $z, w \in \mathbb{C}$. Then there are $a, b, c, d \in \mathbb{R}$ such that $z = a + bi$ and $w = c + di$. Since addition is commutative in $\mathbb{R}$, we have

$$z + w = (a + bi) + (c + di) = (a + c) + (b + d)i = (c + a) + (d + b)i$$
$$= (c + di) + (a + bi) = w + z.$$

(Identity) Let $\overline{0} = 0 + 0i$. We show that $\overline{0}$ is an additive identity for $\mathbb{C}$. Since $0 \in \mathbb{R}$, $\overline{0} \in \mathbb{C}$. Let $z \in \mathbb{C}$. Then there are $a, b \in \mathbb{R}$ such that $z = a + bi$. Since 0 is an additive identity in $\mathbb{R}$, we have

$$\overline{0} + z = (0 + 0i) + (a + bi) = (0 + a) + (0 + b)i = a + bi.$$

$$z + \overline{0} = (a + bi) + (0 + 0i) = (a + 0) + (b + 0)i = a + bi.$$

(Inverse) Let $z \in \mathbb{C}$. Then there are $a, b \in \mathbb{R}$ such that $z = a + bi$. Let $w = -a + (-b)i$. Then

$$z + w = (a + bi) + (-a + (-b)i) = \big(a + (-a)\big) + \big(b + (-b)\big)i = 0 + 0i = \overline{0}.$$

$$w + z = (-a + (-b)i) + (a + bi) = (-a + a) + (-b + b)i = 0 + 0i = \overline{0}.$$

We next prove that $(\mathbb{C}^*, \cdot)$ is a commutative group.

(Closure) Let $z, w \in \mathbb{C}^*$. Then there are $a, b, c, d \in \mathbb{R}$ such that $z = a + bi$ and $w = c + di$. By definition, $zw = (a + bi)(c + di) = (ac - bd) + (ad + bc)i$. Since $\mathbb{R}$ is closed under multiplication, we have $ac, bd, ad, bc \in \mathbb{R}$. Also, $-bd$ is the additive inverse of bd in $\mathbb{R}$. Since $\mathbb{R}$ is closed under addition, we have $ac - bd = ac + (-bd) \in \mathbb{R}$ and $ad + bc \in \mathbb{R}$. Therefore, $zw \in \mathbb{C}$.

We still need to show that $zw \neq 0$. If $zw = 0$, then $ac - bd = 0$ and $ad + bc = 0$. So, $ac = bd$ and $ad = -bc$. Multiplying each side of the last equation by c gives us $acd = -bc^2$. Replacing ac with bd on the left gives $bd^2 = -bc^2$, or equivalently, $bd^2 + bc^2 = 0$. So, $b(d^2 + c^2) = 0$. If $d^2 + c^2 = 0$, then $c = 0$ and $d = 0$, and so, $w = 0$. If $b = 0$, then $ac = 0$, and so, $a = 0$ or $c = 0$. If $a = 0$, then $z = 0$. If $c = 0$ and $a \neq 0$, then since $ad = -bc = 0$, we have $d = 0$. So, $w = 0$. So, we see that $zw = 0$ implies $z = 0$ or $w = 0$. By contrapositive, since $z, w \in \mathbb{C}^*$, we must have $zw \neq 0$, and so, $zw \in \mathbb{C}^*$.

(Associativity) Let $z, w, v \in \mathbb{C}^*$. Then there are $a, b, c, d, e, f \in \mathbb{R}$ such that $z = a + bi$, $w = c + di$, and $v = e + fi$. Since addition and multiplication are associative in $\mathbb{R}$, addition is commutative in $\mathbb{R}$, and multiplication is distributive over addition in $\mathbb{R}$, we have

$$(zw)v = \big((a + bi)(c + di)\big)(e + fi) = \big((ac - bd) + (ad + bc)i\big)(e + fi)$$
$$= [(ac - bd)e - (ad + bc)f] + [(ac - bd)f + (ad + bc)e]i$$
$$= (ace - bde - adf - bcf) + (acf - bdf + ade + bce)i$$
$$= (ace - adf - bcf - bde) + (acf + ade + bce - bdf)i$$
$$= [a(ce - df) - b(cf + de)] + [a(cf + de) + b(ce - df)]i$$
$$= (a + bi)\big((ce - df) + (cf + de)i\big) = (a + bi)\big((c + di)(e + fi)\big) = z(wv).$$

(Commutativity) Let $z, w \in \mathbb{C}^*$. Then there are $a, b, c, d \in \mathbb{R}$ such that $z = a + bi$ and $w = c + di$. Since addition and multiplication are commutative in $\mathbb{R}$, we have

$$zw = (a + bi)(c + di) = (ac - bd) + (ad + bc)i$$
$$= (ca - db) + (cb + da)i = (c + di)(a + bi) = wz$$

(Identity) Let $\overline{1} = 1 + 0i$. We show that $\overline{1}$ is a multiplicative identity for $\mathbb{C}^*$. Since $0, 1 \in \mathbb{R}$, $\overline{1} \in \mathbb{C}^*$. Let $z \in \mathbb{C}^*$. Then there are $a, b \in \mathbb{R}$ such that $z = a + bi$. Since 0 is an additive identity in $\mathbb{R}$, 1 is a multiplicative identity in $\mathbb{R}$, and $0 \cdot x = x \cdot 0 = 0$ for all $x \in \mathbb{R}$, we have

$$\overline{1}z = (1 + 0i)(a + bi) = (1a - 0b) + (1b + 0a)i = 1a + 1bi = a + bi.$$

$$z \cdot \overline{1} = (a + bi)(1 + 0i) = (a \cdot 1 - b \cdot 0) + (a \cdot 0 + b \cdot 1)i = a \cdot 1 + b \cdot 1i = a + bi.$$

(Inverse) Let $z \in \mathbb{C}^*$. Then there are $a, b \in \mathbb{R}$ such that $z = a + bi$. Let $w = \frac{a}{a^2+b^2} + \frac{-b}{a^2+b^2}i$. Then we have

$$zw = (a + bi)\left(\frac{a}{a^2 + b^2} + \frac{-b}{a^2 + b^2}i\right)$$

$$= \left(a \cdot \frac{a}{a^2 + b^2} - b \cdot \frac{-b}{a^2 + b^2}\right) + \left(a \cdot \frac{-b}{a^2 + b^2} + b \cdot \frac{a}{a^2 + b^2}\right)i$$

$$= \frac{a^2 + b^2}{a^2 + b^2} + \frac{-ab + ba}{a^2 + b^2}i = 1 + 0i = \overline{1}.$$

$$wz = \left(\frac{a}{a^2 + b^2} + \frac{-b}{a^2 + b^2}i\right)(a + bi)$$

$$= \left(\frac{a}{a^2 + b^2} \cdot a - \frac{-b}{a^2 + b^2} \cdot b\right) + \left(\frac{a}{a^2 + b^2} \cdot b + \frac{-b}{a^2 + b^2} \cdot a\right)i$$

$$= \frac{a^2 + b^2}{a^2 + b^2} + \frac{ab - ba}{a^2 + b^2}i = 1 + 0i = \overline{1}.$$

(Left Distributivity) Let $z, w, v \in \mathbb{C}$. Then there are $a, b, c, d, e, f \in \mathbb{R}$ such that $z = a + bi$, $w = c + di$, and $v = e + fi$. Since multiplication is left distributive over addition in $\mathbb{R}$, and addition is associative and commutative in $\mathbb{R}$, we have

$$z(w + v) = (a + bi)[(c + di) + (e + fi)] = (a + bi)[(c + e) + (d + f)i]$$

$$= [a(c + e) - b(d + f)] + [a(d + f) + b(c + e)]i$$

$$= (ac + ae - bd - bf) + (ad + af + bc + be)i$$

$$= [(ac - bd) + (ad + bc)i\,] + [(ae - bf) + (af + be)i]$$

$$(a + bi)(c + di) + (a + bi)(e + fi) = zw + zv.$$

(Right Distributivity) Let $z, w, v \in \mathbb{C}$. There are $a, b, c, d, e, f \in \mathbb{R}$ such that $z = a + bi$, $w = c + di$, and $v = e + fi$. Since multiplication is right distributive over addition in $\mathbb{R}$, and addition is associative and commutative in $\mathbb{R}$, we have

$$(w + v)z = [(c + di) + (e + fi)](a + bi) = [(c + e) + (d + f)i](a + bi)$$

$$= [(c + e)a - (d + f)b] + [(c + e)b + (d + f)a]i$$

$$= (ca + ea - db - fb) + (cb + eb + da + fa)i$$

$$= [(ca - db) + (cb + da)i\,] + [(ea - fb) + (eb + fa)i]$$

$$(c + di)(a + bi) + (e + fi)(a + bi) = wz + vz.$$

Therefore, $(\mathbb{C}, +, \cdot)$ is field.

By part 3 of Example 6.5, $(\mathbb{C}, +, \cdot)$ cannot be ordered. □

Note: When verifying the inverse property, we didn't mention the field properties that were used and we skipped some steps. The dedicated reader may want to fill in these details.

18. Prove that every nonempty set of real numbers that is bounded below has a greatest lower bound in $\mathbb{R}$.

Proof: Let S be a nonempty set of real numbers that is bounded below. Let K be a lower bound for S, so that for all $x \in S$, $x \geq K$. Define the set T by $T = \{-x \mid x \in S\}$.

Let $y \in T$. Then there is $x \in S$ with $y = -x$. Since $x \in S$, $x \geq K$. It follows from Problem 11, part (ii) that $y = -x \leq -K$. Since $y \in T$ was arbitrary, we have shown that for all $y \in T$, $y \leq -K$. It follows that $-K$ is an upper bound for the set T.

By the Completeness Property of $\mathbb{R}$, T has a least upper bound M. We will show that $-M$ is a greatest lower bound for S.

Let $x \in S$. Then $-x \in T$. Since M is an upper bound for T, $-x \leq M$. So, by Problem 11, part (ii), $x \geq -M$. Since $x \in S$ was arbitrary, we have shown that for all $x \in S$, $x \geq -M$. Therefore, $-M$ is a lower bound for S.

Let $B > -M$. Then $-B < M$. Since M is the least upper bound for T, there is $y \in T$ with $y > -B$. By Problem 11, part (ii), we have $-y < B$. Since $y \in T$, $-y \in S$. Thus, B is not a lower bound of S.

Therefore, $-M$ is a greatest lower bound for S.

Since S was arbitrary, we have shown that every nonempty set of real numbers that is bounded below has a greatest lower bound in $\mathbb{R}$. □

19. Show that between any two real numbers there is a real number that is **not** rational.

Proof: Let $x, y \in \mathbb{R}$ with $x < y$. Let c be a positive number that is not rational. Then $\frac{x}{c} < \frac{y}{c}$. By the Density Theorem, there is a $q \in \mathbb{Q}$ such that $\frac{x}{c} < q < \frac{y}{c}$. We can assume that $q \neq 0$ (if it were, we could simply apply the Density Theorem again to get $p \in \mathbb{Q}$ with $\frac{x}{c} < p < q$, and p would not be 0). It follows that $x < cq < y$. Since $c = (cq)q^{-1}$, it follows that $cq \notin \mathbb{Q}$ (if $cq \in \mathbb{Q}$, then $c \in \mathbb{Q}$ because $\mathbb{Q}$ is closed under multiplication). So, cq is a real number between x and y that is **not** rational. □

20. Let $T = \{x \in F \mid -2 < x \leq 2\}$. Prove $\sup T = 2$ and $\inf T = -2$.

Proof: If $x \in T$, then by the definition of T, $x \leq 2$. So, 2 is an upper bound of T.

Now, let $B < 2$, and let $z = \max\left\{0, \frac{1}{2}(B+2)\right\}$. Since $B < 2$, we have

$$\frac{1}{2}(B+2) < \frac{1}{2}(2+2) = \frac{1}{2} \cdot 4 = 2.$$

So, if we have $\frac{1}{2}(B+2) > 0$, then $\frac{1}{2}(B+2) \in T$. Since $0 \in T$, we see that $z \in T$. Also,

$$z \geq \frac{1}{2}(B+2) > \frac{1}{2}(B+B) = \frac{1}{2}(2B) = \left(\frac{1}{2} \cdot 2\right)B = 1B = B.$$

So, we see that $z \in T$ and $z > B$. Therefore, B is not an upper bound of T. So, $2 = \sup T$.

If $x \in T$, then by the definition of T, $x > -2$. So, -2 is a lower bound of T.

Now, let $C > -2$, and let $w = \min\left\{0, \frac{1}{2}(-2+C)\right\}$. Since $C > -2$, we have

$$\frac{1}{2}(-2+C) > \frac{1}{2}(-2-2) = \frac{1}{2}(-4) = -2.$$

So, if we have $\frac{1}{2}(-2+C) < 0$, then $\frac{1}{2}(-2+C) \in T$. Since $0 \in T$, we see that $w \in T$. Also,

$$w \leq \frac{1}{2}(-2+C) < \frac{1}{2}(C+C) = \frac{1}{2}(2C) = \left(\frac{1}{2} \cdot 2\right)C = 1C = C.$$

So, we see that $w \in T$ and $w < C$. Therefore, C is not a lower bound of T. So, $-2 = \inf T$. □

Problem Set 7

LEVEL 1

1. Let $z = -4 - i$ and $w = 3 - 5i$. Compute the distance between z and w.

Solution: $|z - w| = |(-4 - i) - (3 - 5i)| = |(-4 - 3) + (-1 + 5)i| = |-7 + 4i| = \sqrt{(-7)^2 + 4^2}$

$$= \sqrt{49 + 16} = \sqrt{\mathbf{65}}.$$

2. Define a set of real numbers with exactly two accumulation points.

Solution: Let $S = \left\{(-1)^n \left(1 + \frac{1}{n}\right) \middle| \, n \in \mathbb{N}\right\}$. Then S has exactly two accumulation points: 1 and -1.

LEVEL 2

3. Let S be a set of real numbers and let S' be the set of accumulation points of S. Prove that S' is closed in $\mathbb{R}$.

Proof: Let S be a set of real numbers, let S' be the set of accumulation points of S, and let x be an accumulation point of S'. We will show that x is an accumulation point of S. To see this, let (a, b) be an open interval containing x. Since x is an accumulation point of S', there is $y \in S'$ with $y \in (a, b)$ and $y \neq x$. Without loss of generality, assume that $x < y$. Since $y \in S'$, y is an accumulation point of S. So, there is $z \in S$ with $z \in (x, b)$. Thus, we have found an element of S in (a, b) different from x. So, x is an accumulation point of S, and therefore, $x \in S'$. Since x was an arbitrary accumulation point of S', it follows that S' contains each of its accumulation points. By Theorem 7.24, S' is closed in $\mathbb{R}$. □

4. Determine the accumulation points of each of the following subsets of $\mathbb{C}$:
   (i) $\left\{\frac{1}{n} \middle| \, n \in \mathbb{Z}^+\right\}$
   (ii) $\left\{\frac{i}{n} \middle| \, n \in \mathbb{Z}^+\right\}$
   (iii) $\{i^n \mid n \in \mathbb{Z}^+\}$
   (iv) $\{z \mid |z| < 1\}$
   (v) $\{z \mid 0 < |z - 2| \leq 3\}$

Solutions:

(i) $\mathbf{0}$ is the only accumulation point of this set.

(ii) $\mathbf{0}$ is the only accumulation point of this set.

(iii) This set is equal to $\{1, -1, i, -i\}$. It has **no accumulation points**.

(iv) The set of accumulation points of the set $\{z \mid |z| < 1\}$ is the set $\{\mathbf{z} \mid |\mathbf{z}| \leq \mathbf{1}\}$.

(v) The set of accumulation points of the set $\{z \mid 0 < |z - 2| \leq 3\}$ is the set $\{\mathbf{z} \mid |\mathbf{z} - \mathbf{2}| \leq \mathbf{3}\}$.

LEVEL 3

5. Prove the following:

   (i) For all $b \in \mathbb{R}$, the infinite interval $(-\infty, b)$ is open in $\mathbb{R}$.

   (ii) The intersection of two open intervals in $\mathbb{R}$ is either empty or an open interval in $\mathbb{R}$.

   (iii) The intersection of finitely many open sets in $\mathbb{R}$ is open in $\mathbb{R}$.

Proofs:

(i) Let $x \in (-\infty, b)$ and let $a = x - 1$. Since $x \in (-\infty, b)$, $x < b$. Since $x - (x - 1) = 1 > 0$, we have $x > x - 1 = a$. So, we have $a < x < b$. That is, $x \in (a, b)$. Also, $(a, b) \subseteq (-\infty, b)$. Since $x \in (-\infty, b)$ was arbitrary, $(-\infty, b)$ is an open set. □

(ii) Let (a, b) and (c, d) be open intervals in $\mathbb{R}$ (a and c can be $-\infty$, and b and d can be ∞, where $-\infty$ is less than any real number and ∞, and ∞ is greater than any real number and $-\infty$). Without loss of generality, we may assume that $a \leq c$. If $b \leq c$, then we have $(a, b) \cap (c, d) = \emptyset$ because if $a < x < b$ and $c < x < d$, then $x < b \leq c < x$, and so, $x < x$, which is impossible.

So, we may assume that $c < b$. Let $e = \min\{b, d\}$. We claim that $(a, b) \cap (c, d) = (c, e)$.

Let $x \in (a, b) \cap (c, d)$. Then $x \in (a, b)$ and $x \in (c, d)$. So, $a < x < b$ and $c < x < d$. In particular, $x > c$, $x < b$, and $x < d$. Since $x < b$ and $x < d$, $x < e$. So, $x \in (c, e)$. Since $x \in (a, b) \cap (c, d)$ was arbitrary, we have shown that $(a, b) \cap (c, d) \subseteq (c, e)$.

Now, let $x \in (c, e)$. Then $c < x < e$. We are assuming that $a \leq c$. We also have $e \leq b$. So, $a \leq c < x < e \leq b$. Therefore, $x \in (a, b)$. We also have $e \leq d$. So, $c < x < e \leq d$, and therefore, $x \in (c, d)$. Since $x \in (a, b)$ and $x \in (c, d)$, we have $x \in (a, b) \cap (c, d)$. Since $x \in (c, e)$ was arbitrary, we have shown that $(c, e) \subseteq (a, b) \cap (c, d)$.

Finally, since we have shown $(a, b) \cap (c, d) \subseteq (c, e)$ and $(c, e) \subseteq (a, b) \cap (c, d)$, we have $(a, b) \cap (c, d) = (c, e)$.

Therefore, the intersection of two open intervals in $\mathbb{R}$ is either empty or an open interval in $\mathbb{R}$. □

(iii) The intersection of a single set with itself is just that set itself, and so, the result holds trivially for one open set.

So, we will prove the following statement: "The intersection of a set of finitely many open sets in $\mathbb{R}$ consisting of at least 2 sets is an open set in $\mathbb{R}$." We will prove this by induction on the number of open sets we are taking the intersection of. Theorem 7.19 is the base case $n = 2$.

For the inductive step, assume that the intersection of k nonempty open sets in $\mathbb{R}$ is open, and let $\boldsymbol{X}$ be a set of $k + 1$ open sets. Let $A \in \boldsymbol{X}$ and let B be the intersection of all the sets in $\boldsymbol{X}$ except A. By the induction hypotheses, B is open. Therefore, $\bigcap \boldsymbol{X} = A \cap B$ is open by Theorem 7.19.

By the Principle of Mathematical Induction, we have shown that the intersection of a set of finitely many open sets in $\mathbb{R}$ consisting of at least 2 sets is an open set in $\mathbb{R}$. □

6. Prove the Triangle Inequality (Theorem 7.4).

Proof: $|z+w|^2 = (z+w)\overline{(z+w)} = (z+w)(\overline{z}+\overline{w}) = z\overline{z} + z\overline{w} + w\overline{z} + w\overline{w}$

$$= z\overline{z} + z\overline{w} + \overline{z\overline{w}} + w\overline{w} = z\overline{z} + 2\mathrm{Re}(z\overline{w}) + w\overline{w} \leq z\overline{z} + 2|z\overline{w}| + w\overline{w} = z\overline{z} + 2|z||\overline{w}| + w\overline{w}$$
$$= |z|^2 + 2|z||w| + |w|^2 = (|z|+|w|)^2$$

Since $|z+w|$ and $|z|+|w|$ are nonnegative, $|z+w| \leq |z|+|w|$. □

7. Let z and w be complex numbers. Prove $\big||z|-|w|\big| \leq |z \pm w| \leq |z|+|w|$.

Proof: $|z| = |(z+w)+(-w)| \leq |z+w| + |-w| = |z+w| + |w|$. So, $|z+w| \geq |z|-|w|$.

$|w| = |(z+w)+(-z)| \leq |z+w| + |-z| = |z+w| + |z|$. So, $|z+w| \geq |w|-|z| = -(|z|-|w|)$.

Since for all $w, z \in \mathbb{C}$, we have $\big||z|-|w|\big| = |z|-|w|$ or $\big||z|-|w|\big| = -(|z|-|w|)$, it follows that $\big||z|-|w|\big| \leq |z+w|$.

Combining this result with the Triangle Inequality, gives us $\big||z|-|w|\big| \leq |z+w| \leq |z|+|w|$.

Now, by the Triangle Inequality we have $|z-w| = |z+(-w)| \leq |z|+|-w| = |z|+|w|$.

Finally, by the third paragraph, we have $|z-w| = |z+(-w)| \geq \big||z|-|-w|\big| = \big||z|-|w|\big|$. □

8. Let $x \in \mathbb{R}$ be an accumulation point of a set S. Prove that every open interval containing x contains infinitely many points of S.

Proof: Let $x \in \mathbb{R}$ be an accumulation point of a set S, let (a,b) be an open interval containing x, and let $X = \{x_1, x_2, \dots, x_n\}$ be an arbitrary subset of $(a,b) \cap S \cap (\mathbb{R} \setminus \{x\})$. Let $A = \{x_i \mid x_i < x\} \cup \{a\}$ and let $B = \{x_i \mid x_i > x\} \cup \{b\}$. Now, let c be the largest element of A and let d be the smallest element of B. Then $c < x < d$, and so, (c,d) is an open interval containing x. Since x is an accumulation point of S, there is $y \in (c,d)$ with $y \in S$ and $y \neq x$. Since $a \leq c$ and $d \leq b$, $(c,d) \subseteq (a,b)$. Therefore, $y \in (a,b)$. Also, $y \notin X$. So, given any finite collection of elements of $Y = (a,b) \cap S \cap (\mathbb{R} \setminus \{x\})$, we can always find another element Y that is not in that finite collection. □

9. Let X, Y, and Z be sets of real numbers with $Z = X \cup Y$. Prove that $\overline{Z} = \overline{X} \cup \overline{Y}$.

Proof: Let X, Y, and Z be sets of real numbers with $Z = X \cup Y$. Let $x \in \overline{Z}$. By part 3 of Theorem 7.25, $x \in Z$ or x is an accumulation point of Z. If $x \in Z$, then $x \in X$ or $x \in Y$. By part 1 of Theorem 7.25, $X \subseteq \overline{X}$ and $Y \subseteq \overline{Y}$. So, $x \in \overline{X}$ or $x \in \overline{Y}$. Thus, $x \in \overline{X} \cup \overline{Y}$. Now, suppose that x is an accumulation point of Z, but x is **not** an accumulation point of Y. Then there is an open interval (c, d) with $x \in (c, d)$ and such that if $y \in Y$ with $y \neq x$, then $y \notin (c, d)$. Let (a, b) be any open interval containing x and let $U = (a, b) \cap (c, d)$. Since $x \in U$, U is a nonempty open interval. Since x is an accumulation point of Z, there is $y \in Z$ with $y \in U$ and $y \neq x$. Since $U \subseteq (c, d)$, $y \notin Y$. Therefore, $y \in X$. Since $U \subseteq (a, b)$, we have $y \in (a, b)$. It follows that x is an accumulation point of X. So, $x \in \overline{X}$. Therefore, $x \in \overline{X}$ or $x \in \overline{Y}$, and it follows that $x \in \overline{X} \cup \overline{Y}$. Since $x \in \overline{Z}$ was arbitrary, we have $\overline{Z} \subseteq \overline{X} \cup \overline{Y}$.

Conversely, let $x \in \overline{X} \cup \overline{Y}$. Then $x \in \overline{X}$ or $x \in \overline{Y}$. Without loss of generality, assume that $x \in \overline{X}$. By part 3 of Theorem 7.25, $x \in X$ or x is an accumulation point of X. If $x \in X$, then $x \in X$ or $x \in Y$. So, $x \in X \cup Y = Z$. By part 1 of Theorem 7.25, $Z \subseteq \overline{Z}$. So, $x \in \overline{Z}$. Now, suppose that x is an accumulation point of X and let (a, b) be any open interval containing x. Then there is $y \in X$ with $y \in (a, b)$ and $y \neq x$. So, $y \in X$ or $y \in Y$, and thus, $y \in X \cup Y = Z$. It follows that x is an accumulation point of Z. Therefore, $x \in \overline{Z}$. Since $x \in \overline{X} \cup \overline{Y}$ was arbitrary, $\overline{X} \cup \overline{Y} \subseteq \overline{Z}$.

Since $\overline{Z} \subseteq \overline{X} \cup \overline{Y}$ and $\overline{X} \cup \overline{Y} \subseteq \overline{Z}$, we have $\overline{Z} = \overline{X} \cup \overline{Y}$. □

10. Prove that if $\boldsymbol{X}$ is a nonempty set of closed subsets of $\mathbb{R}$, then $\bigcap \boldsymbol{X}$ is closed.

Proof: Let $\boldsymbol{X}$ be a nonempty set of closed subsets of $\mathbb{R}$. Then for each $A \in \boldsymbol{X}$, $\mathbb{R} \setminus A$ is an open set in $\mathbb{R}$. By Theorem 7.15, $\bigcup\{\mathbb{R} \setminus A \mid A \in \boldsymbol{X}\}$ is open in $\mathbb{R}$. Therefore, $\mathbb{R} \setminus \bigcup\{\mathbb{R} \setminus A \mid A \in \boldsymbol{X}\}$ is closed in $\mathbb{R}$. So, it suffices to show that $\bigcap \boldsymbol{X} = \mathbb{R} \setminus \bigcup\{\mathbb{R} \setminus A \mid A \in \boldsymbol{X}\}$. Well, $x \in \bigcap \boldsymbol{X}$ if and only if for all $A \in \boldsymbol{X}$, $x \in A$ if and only if for all $A \in \boldsymbol{X}$, $x \notin \mathbb{R} \setminus A$ if and only if $x \notin \bigcup\{\mathbb{R} \setminus A \mid A \in \boldsymbol{X}\}$ if and only if $x \in \mathbb{R} \setminus \bigcup\{\mathbb{R} \setminus A \mid A \in \boldsymbol{X}\}$. So, $\bigcap \boldsymbol{X} = \mathbb{R} \setminus \bigcup\{\mathbb{R} \setminus A \mid A \in \boldsymbol{X}\}$, completing the proof. □

LEVEL 4

11. Give an example of an infinite collection of open sets in $\mathbb{R}$ whose intersection is not open in $\mathbb{R}$. Also, give an example of an infinite collection of closed sets in $\mathbb{R}$ whose union is not closed in $\mathbb{R}$. Provide a proof for each example.

Solution: Let $\boldsymbol{X} = \left\{\left(0, 1 + \frac{1}{n}\right) \mid n \in \mathbb{Z}^+\right\}$. Each set in $\boldsymbol{X}$ is an open interval, and therefore, open. We will show that $\bigcap \boldsymbol{X} = (0,1]$. Note that $x \in \bigcap \boldsymbol{X}$ if and only if for all $n \in \mathbb{Z}^+$, $x \in \left(0, 1 + \frac{1}{n}\right)$ if and only if for all $n \in \mathbb{Z}^+$, $0 < x < 1 + \frac{1}{n}$. We need to show that $x \leq 1$ is equivalent to $\forall n \in \mathbb{Z}^+ \left(x < 1 + \frac{1}{n}\right)$.

Suppose that $x \leq 1$. Let $n \in \mathbb{Z}^+$. By Theorem 6.8, $\frac{1}{n} > 0$. So, $1 + \frac{1}{n} - 1 > 0$ (SACT). Thus, $1 + \frac{1}{n} > 1$. So, we have $x \leq 1 < 1 + \frac{1}{n}$, and therefore, $x < 1 + \frac{1}{n}$. Since $n \in \mathbb{Z}^+$ was arbitrary, we have shown that $x \leq 1$ implies $\forall n \in \mathbb{Z}^+ \left(x < 1 + \frac{1}{n}\right)$.

Now, suppose $x > 1$ (proof by contrapositive). Then $x - 1 > 0$. Since there is no smallest positive real number, there is a real number $\epsilon > 0$ with $x - 1 > \epsilon$. By the Archimedean Property of the reals, there is a natural number n with $n > \frac{1}{\epsilon}$. So, $\frac{1}{n} < \epsilon$, or equivalently, $\epsilon > \frac{1}{n}$. Thus, $x - 1 > \frac{1}{n}$, and so, $x > 1 + \frac{1}{n}$. We have shown that there is $n \in \mathbb{Z}^+$ such that $x \geq 1 + \frac{1}{n}$. So, $\forall n \in \mathbb{Z}^+ \left(x < 1 + \frac{1}{n}\right)$ is false.

This equivalence proves that $\bigcap \boldsymbol{X} = (0,1]$, an interval that is **not** an open set.

Let $\boldsymbol{Y} = \left\{\left[0, 1 - \frac{1}{n}\right] \mid n \in \mathbb{Z}^+\right\}$. Each set in Y is a closed interval, and therefore, closed. We will show that $\bigcup \boldsymbol{Y} = [0,1)$. Note that $x \in \bigcup Y$ if and only if there is $n \in \mathbb{Z}^+$ such that $x \in \left[0, 1 - \frac{1}{n}\right]$ if and only if there is $n \in \mathbb{Z}^+$ such that $0 \leq x \leq 1 - \frac{1}{n}$. We need to show that $x < 1$ is equivalent to $\exists n \in \mathbb{Z}^+ \left(x \leq 1 - \frac{1}{n}\right)$ (where $\exists$ is read "there exists" or "there is").

Suppose $x < 1$. Then $1 - x > 0$. Since there is no smallest positive real number, there is a real number $\epsilon > 0$ with $1 - x > \epsilon$. By the Archimedean Property of the reals, there is a natural number n with $n > \frac{1}{\epsilon}$. So, $\frac{1}{n} < \epsilon$, or equivalently, $\epsilon > \frac{1}{n}$. Thus, $1 - x > \frac{1}{n}$, and so, $x < 1 - \frac{1}{n}$. We have shown that there is $n \in \mathbb{Z}^+$ such that $x \leq 1 - \frac{1}{n}$. So, $\exists n \in \mathbb{Z}^+ \left(x \leq 1 - \frac{1}{n}\right)$.

Now, suppose $x \geq 1$ (proof by contrapositive). Let $n \in \mathbb{Z}^+$. By Theorem 6.8, $\frac{1}{n} > 0$. So, $1 - 1 + \frac{1}{n} > 0$ (SACT). So, $1 > 1 - \frac{1}{n}$. It follows that $x > 1 - \frac{1}{n}$. Since $n \in \mathbb{Z}^+$ was arbitrary, $\forall n \in \mathbb{Z}^+ \left(x > 1 - \frac{1}{n}\right)$. It follows that $\exists n \in \mathbb{Z}^+ \left(x \leq 1 - \frac{1}{n}\right)$ is false.

This equivalence proves that $\bigcup Y = [0,1)$, an interval that is **not** a closed set.

12. Determine if each of the following subsets of $\mathbb{C}$ is open, closed, both, or neither. Give a proof in each case.

    (i) $\emptyset$

    (ii) $\mathbb{C}$

    (iii) $\{z \in \mathbb{C} \mid |z| > 1\}$

    (iv) $\{z \in \mathbb{C} \mid \text{Im } z \leq -2\}$

    (v) $\{i^n \mid n \in \mathbb{Z}^+\}$

    (vi) $\{z \in \mathbb{C} \mid 2 < |z - 2| < 4\}$

Proofs:

(i) $\emptyset$ is **open and closed**. The statement that $\emptyset$ is open is vacuously true (since $\emptyset$ has no elements, there is nothing to check). $\emptyset$ is closed because $\mathbb{C} \setminus \emptyset = \mathbb{C}$ is open (proof below). □

(ii) $\mathbb{C}$ is **open and closed**. $\mathbb{C}$ is closed because $\mathbb{C} \setminus \mathbb{C} = \emptyset$ is open (see (i)). To see that $\mathbb{C}$ is open. Let $a \in \mathbb{C}$, and let $D = \{z \in \mathbb{C} \mid |z - a| < 1\}$ be the open disk with center a and radius 1. Since $|a - a| = 0 < 1$, $a \in D$, and since every element of D is a complex number, $D \subseteq \mathbb{C}$. It follows that $\mathbb{C}$ is open. □

(iii) $S = \{z \in \mathbb{C} \mid |z| > 1\}$ is **open**. To see this, let $w \in S$ and let $r = |w| - 1$. We will show that $N_r(w) \subseteq S$ (recall that $N_r(w) = \{z \in \mathbb{C} \mid |z - w| < r\}$). Let $z \in N_r(w)$. Then we have $|z - w| < r = |w| - 1$. So, using the Triangle Inequality, we have

$$|w| = |(w - z) + z| \leq |w - z| + |z| = |z - w| + |z| < |w| - 1 + |z|$$

Thus, $|w| < |w| - 1 + |z|$, and therefore, $|z| > 1$. So, $z \in S$. Since $z \in N_r(w)$ was arbitrary, we have shown that $N_r(w) \subseteq S$. So, S is open. □

$S = \{z \in \mathbb{C} \mid |z| > 1\}$ is **not closed** because $\mathbb{C} \setminus S = \{z \in \mathbb{C} \mid |z| \leq 1\}$ is not open.

(iv) $S = \{z \in \mathbb{C} \mid \operatorname{Im} z \leq -2\}$ is **not open**. To see this, first note that $\operatorname{Im}(-2i) = -2$, and so, $-2i \in S$. If $N_r(-2i)$ is an r-neighborhood of $-2i$, then $\frac{-4+r}{2} i \in N_r(-2i)$ because $\left|\frac{-4+r}{2} i - (-2i)\right| = \left|\left(\frac{-4+r}{2} + \frac{4}{2}\right) i\right| = \left|\frac{r}{2}\right| = \frac{r}{2} < r$. $\frac{-4+r}{2} i \notin S$ because $\operatorname{Im}\left(\frac{-4+r}{2} i\right) = \frac{-4+r}{2}$, and $\frac{-4+r}{2} > -\frac{4}{2} = -2$. So, for all $r > 0$, $N_r(-2i) \nsubseteq S$, showing that S is not open. □

S is **closed**. To see this, we show that $T = \mathbb{C} \setminus S = \{z \in \mathbb{C} \mid \operatorname{Im} z > -2\}$ is open.

Let $w \in T$ and let $r = 2 + \operatorname{Im} w$. We will show that $N_r(w) \subseteq T$. Let $z \in N_r(w)$. Then we have $\operatorname{Im} w - \operatorname{Im} z = \operatorname{Im}(w - z) \leq |w - z| = |z - w| < r = 2 + \operatorname{Im} w$. So, $-\operatorname{Im} z < 2$, and therefore, $\operatorname{Im} z > -2$. So, $z \in T$. Since $z \in N_r(w)$ was arbitrary, we have shown that $N_r(w) \subseteq T$. So, T is open, and therefore, $S = \mathbb{C} \setminus T$ is closed. □

(v) Note that $S = \{i^n \mid n \in \mathbb{Z}^+\}$ is a finite set consisting of just four complex numbers. Indeed, $S = \{1, -1, i, -i\}$.

S is **not open**. To see this, let $N_r(i)$ be an arbitrary r-neighborhood of i. Then $i + \frac{r}{2} \in N_r(i)$ because $\left|\left(i + \frac{r}{2}\right) - i\right| = \left|\frac{r}{2}\right| = \frac{r}{2} < r$, but $i + \frac{r}{2} \notin S$ because $i + \frac{r}{2}$ is not equal to $1, -1, i$, or $-i$. □

S is **closed**. To see this, we show that $T = \mathbb{C} \setminus S$ is open.

Let $w \in T$ and let $r = \min\{|w - 1|, |w + 1|, |w - i|, |w + i|\}$.

We now show that $N_r(w) \subseteq T$. Since $r \leq |w - 1|$, $1 \notin N_r(w)$ (otherwise, $|w - 1| < r$). Similarly, $-1, i$, and $-i \notin N_r(w)$. So, if $z \in N_r(w)$, then $z \notin S$, and so, $z \in T$.

By Theorem 7.28, $\mathbb{C} \setminus S$ is open. Therefore, S is closed. □

(vi) $S = \{z \in \mathbb{C} \mid 2 < |z - 2| < 4\}$ is **open** and **not closed**.

To see that S is open, let $z \in S$ and let $r = \min\{|z - 2| - 2, 4 - |z - 2|\}$. We show that $D_r(z) \subseteq S$. Let $w \in D_r(z)$. Then $|w - z| < r$. So, $|w - z| < 4 - |z - 2|$. Therefore, we have $|w - z| + |z - 2| < 4$, and so, $|w - 2| = |w - z + z - 2| \leq |w - z| + |z - 2| < 4$.

Also, $|w - z| < |z - 2| - 2$. So, we have

$$2 < |z - 2| - |w - z| = |z - w + w - 2| - |w - z| \leq |z - w| + |w - 2| - |w - z|$$
$$= |w - z| + |w - 2| - |w - z| = |w - 2|.$$

So, $2 < |w - 2| < 4$, and therefore, $w \in S$.

Since $z \in S$ was arbitrary, S is open.

To see that S is not closed, we show that $\mathbb{C} \setminus S = \{z \in \mathbb{C} \mid |z-2| \leq 2 \text{ or } |z-2| \geq 4\}$ is not open. To see this, first note that $|6-2| = |4| = 4 \geq 4$, and so, $6 \in \mathbb{C} \setminus S$. Let $N_r(6)$ be an r-neighborhood of 6 and let $k = \min\left\{1, \frac{r}{2}\right\}$ then we have $6 - k \in N_r(6)$ because $|(6-k)-6| = |-k| = k \leq \frac{r}{2} < r$. However, $6 - k \notin \mathbb{C} \setminus S$. To see this, first observe that $|(6-k)-2| = |4-k|$. If $k = 1$, then $|4-k| = |4-1| = |3| = 3$ and it follows that $6 - k \notin \mathbb{C} \setminus S$. If $k = \frac{r}{2}$, then $0 < \frac{r}{2} \leq 1$, so that $-1 \leq -\frac{r}{2} < 0$, and thus, $3 < 4 - \frac{r}{2} < 4$. So, $3 < \left|4 - \frac{r}{2}\right| < 4$ and once again, $6 - k \notin \mathbb{C} \setminus S$. So, $\mathbb{C} \setminus S$ is not open. Therefore, S is not closed. □

13. Prove the following:

(i) An arbitrary union of open sets in $\mathbb{C}$ is an open set in $\mathbb{C}$.

(ii) A finite intersection of open sets in $\mathbb{C}$ is an open set in $\mathbb{C}$.

(iii) An arbitrary intersection of closed sets in $\mathbb{C}$ is a closed set in $\mathbb{C}$.

(iv) A finite union of closed sets in $\mathbb{C}$ is a closed set in $\mathbb{C}$.

(v) Every open set in $\mathbb{C}$ can be expressed as a union of open disks.

Proofs:

(i) Let $\boldsymbol{X}$ be a set of open subsets of $\mathbb{C}$. and let $z \in \bigcup \boldsymbol{X}$. Then $z \in A$ for some $A \in \boldsymbol{X}$. Since A is open in $\mathbb{C}$, there is an open disk D with $z \in D$ and $D \subseteq A$. By part (i) of Problem 9 from Problem Set 2, we have $A \subseteq \bigcup \boldsymbol{X}$. Since $\subseteq$ is transitive (Theorem 1.14), $D \subseteq \bigcup \boldsymbol{X}$. Therefore, $\bigcup \boldsymbol{X}$ is open. □

(ii) Let $\boldsymbol{X}$ be a finite set of open sets in $\mathbb{C}$. If $\bigcap \boldsymbol{X} = \emptyset$, then $\bigcap \boldsymbol{X}$ is open by the proof of Problem 12 (part (i)). Otherwise, let $z \in \bigcap \boldsymbol{X}$. By Theorem 7.28, for each A in $\boldsymbol{X}$, there is an open disk D_A with center z and radius r_A such that $z \in D_A$ and $D_A \subseteq A$. Let $r = \min\{r_A \mid A \in \boldsymbol{X}\}$ and let D be the open disk with center z and radius r. Since $D = D_A$ for some $A \in \boldsymbol{X}$, $z \in D$. Let $w \in D$ and let $B \in \boldsymbol{X}$. Then $|z - w| < r \leq r_B$. So, $w \in D_B$. Since $B \in \boldsymbol{X}$ was arbitrary, $w \in \bigcap \boldsymbol{X}$. Therefore, $D \subseteq \bigcap \boldsymbol{X}$, and so $\bigcap \boldsymbol{X}$ is open. □

(iii) Let $\boldsymbol{X}$ be a nonempty set of closed sets in $\mathbb{C}$. Then for each $A \in \boldsymbol{X}$, $\mathbb{C} \setminus A$ is an open set in $\mathbb{C}$. By (i), $\bigcup\{\mathbb{C} \setminus A | A \in \boldsymbol{X}\}$ is open in $\mathbb{C}$. Therefore, $\mathbb{C} \setminus \bigcup\{\mathbb{C} \setminus A | A \in \boldsymbol{X}\}$ is closed in $\mathbb{C}$. So, it suffices to show that $\bigcap \boldsymbol{X} = \mathbb{C} \setminus \bigcup\{\mathbb{C} \setminus A | A \in \boldsymbol{X}\}$. Well, $x \in \bigcap \boldsymbol{X}$ if and only if for all $A \in \boldsymbol{X}$, $x \in A$ if and only if for all $A \in \boldsymbol{X}$, $x \notin \mathbb{C} \setminus A$ if and only if $x \notin \bigcup\{\mathbb{C} \setminus A | A \in \boldsymbol{X}\}$ if and only if $x \in \mathbb{C} \setminus \bigcup\{\mathbb{C} \setminus A | A \in \boldsymbol{X}\}$. So, $\bigcap \boldsymbol{X} = \mathbb{C} \setminus \bigcup\{\mathbb{C} \setminus A | A \in \boldsymbol{X}\}$, completing the proof. □

(iv) Let $\boldsymbol{X}$ be a finite set of closed subsets of $\mathbb{C}$. Then for each $A \in \boldsymbol{X}$, $\mathbb{C} \setminus A$ is an open set in $\mathbb{C}$. By (ii), $\bigcap\{\mathbb{C} \setminus A | A \in \boldsymbol{X}\}$ is open in $\mathbb{C}$. Therefore, $\mathbb{C} \setminus \bigcap\{\mathbb{C} \setminus A | A \in \boldsymbol{X}\}$ is closed in $\mathbb{C}$. So, it suffices to show that $\bigcup \boldsymbol{X} = \mathbb{C} \setminus \bigcap\{\mathbb{C} \setminus A | A \in \boldsymbol{X}\}$. Well, $x \in \bigcup \boldsymbol{X}$ if and only if there is an $A \in \boldsymbol{X}$ such that $x \in A$ if and only if there is an $A \in \boldsymbol{X}$ such that $x \notin \mathbb{C} \setminus A$ if and only if $x \notin \bigcap\{\mathbb{C} \setminus A | A \in \boldsymbol{X}\}$ if and only if $x \in \mathbb{C} \setminus \bigcap\{\mathbb{C} \setminus A | A \in \boldsymbol{X}\}$. So, $\bigcup \boldsymbol{X} = \mathbb{C} \setminus \bigcap\{\mathbb{C} \setminus A | A \in \boldsymbol{X}\}$, completing the proof. □

(v) Let X be an open set in $\mathbb{C}$. Since X is open, for each $z \in X$, there is an open disk D_z with $z \in D_z$ and $D_z \subseteq X$. We Let $Y = \{D_z \mid z \in X\}$. We will show that $X = \bigcup Y$.

First, let $z \in X$. Then $z \in D_z$. Since $D_z \in Y$, $z \in \bigcup Y$. Since z was arbitrary, $X \subseteq \bigcup Y$.

Now, let $z \in \bigcup Y$. Then there is $w \in X$ with $z \in D_w$. Since $D_w \subseteq X$, $z \in X$. Since z was arbitrary, $\bigcup Y \subseteq X$.

Since $X \subseteq \bigcup Y$ and $\bigcup Y \subseteq X$, it follows that $X = \bigcup Y$. □

LEVEL 5

14. Prove that every closed set in $\mathbb{R}$ can be written as an intersection $\bigcap \boldsymbol{X}$, where each element of $\boldsymbol{X}$ is a union of at most 2 closed intervals.

Proof: First note that $\mathbb{R} = \bigcap\{\mathbb{R}\}$.

Let A be a closed set in $\mathbb{R}$ with $A \neq \mathbb{R}$. Then $\mathbb{R} \setminus A$ is a nonempty open set in $\mathbb{R}$. By Theorem 7.17, $\mathbb{R} \setminus A$ can be expressed as $\bigcup \boldsymbol{X}$, where $\boldsymbol{X}$ is a set of bounded open intervals. For each B in $\boldsymbol{X}$, $\mathbb{R} \setminus B$ is a union of two closed intervals (if $B = (a, b)$, then $\mathbb{R} \setminus B = (-\infty, a] \cup [b, \infty)$). Now, by part (iii) of Problem 10 from Problem Set 2, we have $A = \mathbb{R} \setminus (\mathbb{R} \setminus A) = \mathbb{R} \setminus \bigcup \boldsymbol{X} = \bigcap\{\mathbb{R} \setminus B \mid B \in \boldsymbol{X}\}$. □

15. A complex number z is an **interior point** of a set S of complex numbers if there is a neighborhood of z that contains only points in S, whereas w is a **boundary point** of S if each neighborhood of w contains at least one point in S and one point not in S. Prove the following:
    (i) A set of complex numbers is open in $\mathbb{C}$ if and only if each point in S is an interior point of S.
    (ii) A set of complex numbers is open in $\mathbb{C}$ if and only if it contains none of its boundary points.
    (iii) A set of complex numbers is closed in $\mathbb{C}$ if and only if it contains all its boundary points.

Proofs:

(i) Let S be a set of complex numbers. Then S is open if and only if for every complex number $z \in S$, there is an open disk D with $z \in D$ and $D \subseteq S$ if and only if for every complex number $z \in S$, there is a neighborhood of z that contains only points in S if and only if every complex number in S is an interior point of S. □

(ii) Suppose that S is an open set of complex numbers and let $z \in S$. By (i), z is an interior point of S. So, there is a neighborhood of z containing only points of S. So, z is **not** a boundary point of S. Since $z \in S$ was arbitrary, S contains none of its boundary points.

We now prove that if S contains none of its boundary points, then S is open by contrapositive. Suppose S is not open. By (i), there is $z \in S$ such that z is **not** an interior point. Let N be a neighborhood of z. Since $z \in S$, N contains a point in S (namely, z). Since z is not an interior point of S, N contains a point not in S. So, z is a boundary point of S. Therefore, S contains at least one of its boundary points. □

(iii) First note that a complex number z is a boundary point of S if and only if z is a boundary point of $\mathbb{C} \setminus S$ (because $z \in S$ if and only if $z \notin \mathbb{C} \setminus S$, and vice versa).

Let S be a set of complex numbers. Then S is closed if and only if $\mathbb{C} \setminus S$ is open if and only if $\mathbb{C} \setminus S$ contains none of its boundary points (by (ii)) if and only if $S = \mathbb{C} \setminus (\mathbb{C} \setminus S)$ contains all its boundary points. □

16. Let $D = \{z \in \mathbb{C} \mid |z| \leq 1\}$ be the closed unit disk and let S be a subset of D that includes the interior of the disk but is missing at least one point on the bounding circle of the disk. Show that S is not a closed set.

Proof: Let S be a set of complex numbers such that $S \subseteq D$, where $D = \{z \in \mathbb{C} \mid |z| \leq 1\}$, such that S contains $\{z \in \mathbb{C} \mid |z| < 1\}$, but is missing some point w with $|w| = 1$. We will show that w is a boundary point of S. To see this, let N be a neighborhood of w with radius r. w is a point in N that is not in S. We need to find a point in N that is in S.

If $r > 1$, let $z = 0$. Since $0 < 1$, $z \in S$. Also, $|z - w| = |0 - w| = |-w| = |w| = 1 < r$. So, $z \in N$.

If $r \leq 1$, let $z = \frac{2-r}{2}w$. Then $z \in N$ because we have

$$|z - w| = \left|\frac{2-r}{2}w - w\right| = \left|\frac{2-r}{2}w - \frac{2}{2}w\right| = \left|\frac{2-r-2}{2}w\right| = \left|-\frac{r}{2}w\right| = \left|-\frac{r}{2}\right||w| = \frac{r}{2} \cdot 1 = \frac{r}{2} < r.$$

Also, we have $z \in S$ because $|z| = \left|\frac{2-r}{2}w\right| = \left|\frac{2-r}{2}\right||w| = \left|1 - \frac{r}{2}\right|(1) = 1 - \frac{r}{2} < 1$.

So, we have found a boundary point of S that is not in S. By Problem 15, part (iii), S is **not** closed. □

17. Prove that a subset C of $\mathbb{C}$ is closed in $\mathbb{C}$ if and only if C contains each of its accumulation points (this is Theorem 7.31).

Proof: Suppose that S is a closed set of complex numbers and let a be an accumulation point of S. Assume toward contradiction that $a \notin S$, and let N be a neighborhood of a. Since a is an accumulation point, N contains a point in S. Since $a \notin S$, N contains a point not in S (namely, a). So, a is a boundary point of S. Since S is closed, by Problem 15 (part (iii)), $a \in S$, contradicting our assumption that $a \notin S$. So, we must have $a \in S$. Since a was an arbitrary accumulation point of S, we see that S contains all its accumulation points.

Now, suppose that S contains all its accumulation points, and let a be a boundary point of S. Assume toward contradiction that $a \notin S$. Then each neighborhood of a contains a point in S that is not equal to a. So, each deleted neighborhood of a contains a point in S. So, a is an accumulation point of S, and therefore, by our assumption that S contains all its accumulation points, $a \in S$. This contradicts our assumption that $a \notin S$. So, we must have $a \in S$. Since a was an arbitrary boundary point of S, we see that S contains all its boundary points. By Problem 15 (part (iii)), S is closed. □

18. Prove that a set consisting of finitely many complex numbers is a closed set in $\mathbb{C}$. (Hint: Show that a finite set has no accumulation points.)

Proof: Let S be a set consisting of finitely many points. We will show that S has no accumulation points. Let $a \in \mathbb{C}$, and let $r = \min\{|w - a| \mid w \in S \wedge w \neq a\}$. Suppose toward contradiction that the deleted neighborhood $N_r^{\odot}(a) = \{z \mid 0 < |z - a| < r\}$ contains a complex number in S. Let's call this complex number w. Since $w \in S$ and $w \neq a$, by the definition of r, we have $|w - a| \geq r$. Since $w \in N_r^{\odot}(a)$, we have $|w - a| < r$. So, $r \leq |w - a| < r$, and therefore, $r < r$, a contradiction. Therefore, a is not an accumulation point of S. Since $a \in \mathbb{C}$ was arbitrary, we have shown that S has no accumulation points.

Since S has no accumulation points, the statement "S contains all its accumulation points" is vacuously true. By Problem 17, S is closed. □

19. Let C be the Cantor set. Prove the following:

(i) $\frac{1}{4} \in C$.

(ii) C is uncountable.

Proofs:

(i) Recall that $C = \bigcap\{C_n \mid n \in \mathbb{N}\}$. First note that if $x \in C_n$, then $\frac{1}{3}x \in C_{n+1}$. Also, note that $x \in C_n$ if and only if $1 - x \in C_n$. Now, we prove by induction that for all $n \in \mathbb{N}$, $\frac{1}{4} \in C_n$ and $\frac{3}{4} \in C_n$.

Base case ($k = 0$): Clearly $\frac{1}{4}, \frac{3}{4} \in C_0 = [0, 1]$.

Inductive step: Assume that $\frac{1}{4}, \frac{3}{4} \in C_k$. Since $\frac{3}{4} \in C_k$, $\frac{1}{4} = \frac{1}{3} \cdot \frac{3}{4} \in C_{k+1}$. By the symmetry of C_{k+1}, $\frac{3}{4} = 1 - \frac{1}{4} \in C_{k+1}$.

By the Principle of Mathematical Induction, for all $n \in \mathbb{N}$, $\frac{1}{4} \in C_n$ and $\frac{3}{4} \in C_n$.

Therefore, $\frac{1}{4}, \frac{3}{4} \in \bigcap\{C_n \mid n \in \mathbb{N}\} = C$. □

(ii) We define an injection $f\colon [0, 1) \to C$ as follows. For $x \in [0, 1)$, let $x = 0.x_0x_1x_2 \ldots$ be the binary expansion of x (see Note 1 following Example 4.17). For each $n \in \mathbb{N}$, let $z_n = 2x_n$ and consider $z = 0.z_0z_1z_2 \ldots$ as the ternary expansion of a real number. Observe that for all $n \in \mathbb{N}$, $z_n = 0$ or $z_n = 2$. Since $z_0 \neq 1$, $z \notin \left[\frac{1}{3}, \frac{2}{3}\right)$. So $z \in C_1$. Since $z_1 \neq 1$, $z \notin \left[\frac{1}{9}, \frac{2}{9}\right)$ and $z \notin \left[\frac{7}{9}, \frac{8}{9}\right)$. So, $z \in C_2$. By a continuation of this reasoning, we see that for all $n \in \mathbb{N}$, $z \in C_n$. Therefore, $z \in \bigcap\{C_n \mid n \in \mathbb{N}\} = C$. So, if we let $f(x) = z$, then $f\colon [0, 1) \to C$. To see that f is injective, observe that if $x \neq y$, then for some $n \in \mathbb{N}$, $x_n \neq y_n$. Without loss of generality, assume that $x_n = 0$ and $y_n = 1$. So, if $f(x) = z$ and $f(y) = w$, then $z_n = 0$ and $w_n = 2$. It follows that $f(x) = y \neq w = f(y)$. □

20. If $X \subseteq Y$, then X is said to be **dense** in Y if $\overline{X} = Y$. Prove that $\mathbb{C}$ has a countable dense subset. Does $\mathbb{R}^n$ have a countable dense subset for each $n \in \mathbb{N}$?

Proof: Let $S = \{a + bi \mid a, b \in \mathbb{Q}\}$, let $z = c + di \in \mathbb{C}$, and let $N_r(z)$ be an r-neighborhood of z. By the Density Theorem, there is $q \in \mathbb{Q}$ with $c < q < c + r$. So, $q - c < r$, and so, $r^2 - (q - c)^2 > 0$. It follows that $d < d + \sqrt{r^2 - (q - c)^2}$. Once again, by the Density Theorem, there is $p \in \mathbb{Q}$ with $d < p < d + \sqrt{r^2 - (q - c)^2}$. Since $q, p \in \mathbb{Q}$, $q + pi \in S$. Since $c < q$, $q + pi \neq z$. Also,

$$|(q+pi)-(c+di)| = |(q-c)+(p-d)i| = \sqrt{(q-c)^2+(p-d)^2}$$
$$< \sqrt{(q-c)^2+\left(\sqrt{r^2-(q-c)^2}\right)^2} = \sqrt{(q-c)^2+r^2-(q-c)^2} = \sqrt{r^2} = r.$$

So, z is an accumulation point of S. Therefore, $z \in \overline{S}$. Since $z \in \mathbb{C}$ was arbitrary, $\mathbb{C} \subseteq \overline{S}$. It follows that $\overline{S} = \mathbb{C}$, and so, S is a dense subset of $\mathbb{C}$.

Since $\mathbb{Q} \sim \mathbb{N}$, it follows that $\mathbb{Q} \times \mathbb{Q} \sim \mathbb{N} \times \mathbb{N} \sim \mathbb{N}$. Therefore, S is a countable dense subset of $\mathbb{C}$.

Similarly, $\mathbb{Q}^n$ is a countable dense subset of $\mathbb{R}^n$ □

Problem Set 8

LEVEL 1

1. Let $f: \mathbb{R} \to \mathbb{R}$ be defined by $f(x) = 5x - 1$.
   (i) Prove that $\lim_{x \to 3} f(x) = 14$.
   (ii) Prove that f is continuous on $\mathbb{R}$.

Proofs:

(i) Let $\epsilon > 0$ and let $\delta = \frac{\epsilon}{5}$. Suppose that $0 < |x - 3| < \delta$. Then we have

$$|(5x - 1) - 14| = |5x - 15| = 5|x - 3| < 5\delta = 5 \cdot \frac{\epsilon}{5} = \epsilon.$$

So, $\lim_{x \to 3} f(x) = 14$. □

(ii) Let $a \in \mathbb{R}$. We will show that f is continuous at a. Let $\epsilon > 0$, let $\delta = \frac{\epsilon}{5}$, and let $|x - a| < \delta$. Then

$$|(5x - 1) - (5a - 1)| = |5x - 5a| = 5|x - a| < 5\delta = 5 \cdot \frac{\epsilon}{5} = \epsilon.$$

So, f is continuous at a. Since $a \in \mathbb{R}$ was arbitrary, f is continuous on $\mathbb{R}$. □

2. Let $r, c \in \mathbb{R}$ and let $f: \mathbb{R} \to \mathbb{R}$ be defined by $f(x) = c$. Prove that $\lim_{x \to r}[f(x)] = c$.

Proof: Let $\epsilon > 0$ and let $\delta = 1$. If $0 < |x - r| < \delta$, then $|f(x) - c| = |c - c| = |0| = 0 < \epsilon$. Therefore, $\lim_{x \to r}[f(x)] = c$. □

3. Let $A \subseteq \mathbb{R}$, let $f: A \to \mathbb{R}$, let $r, k \in \mathbb{R}$, and suppose that $\lim_{x \to r}[f(x)]$ is a finite real number. Prove that $\lim_{x \to r}[kf(x)] = k \lim_{x \to r}[f(x)]$.

Proof: Suppose that $\lim_{x \to r}[f(x)] = L$ and let $\epsilon > 0$. First assume that $k \neq 0$. Since $\lim_{x \to r}[f(x)] = L$, there is $\delta > 0$ such that $0 < |x - r| < \delta$ implies $|f(x) - L| < \frac{\epsilon}{|k|}$. Suppose that $0 < |x - r| < \delta$. Then

$$|kf(x) - kL| = |k||f(x) - L| < |k| \frac{\epsilon}{|k|} = \epsilon.$$

So, $\lim_{x \to r}[kf(x)] = kL = k \lim_{x \to r}[f(x)]$.

If $k = 0$, let $\delta = 1$. If $0 < |x - r| < \delta$, then

$$|kf(x) - kL| = |0f(x) - 0L| = |0| = 0 < \epsilon.$$

So, in this case, we also have $\lim_{x \to r}[kf(x)] = kL = k \lim_{x \to r}[f(x)]$. □

LEVEL 2

4. Let $A \subseteq \mathbb{R}$, let $f: A \to \mathbb{R}$, and let $r \in \mathbb{R}$. Prove that f is continuous at r if and only if $\lim_{x \to r}[f(x)] = f(r)$.

Proof: Let $A \subseteq \mathbb{R}$, let $f: A \to \mathbb{R}$, and let $r \in \mathbb{R}$. First suppose that f is continuous at r. Let $\epsilon > 0$. Then there is $\delta > 0$ such that $|x - r| < \delta$ implies $|f(x) - f(r)| < \epsilon$. Let $x \in \mathbb{R}$ satisfy $0 < |x - r| < \delta$. Then $|x - r| < \delta$. So, $|f(x) - f(r)| < \epsilon$. Since $\epsilon > 0$ was arbitrary, $\lim_{x \to r}[f(x)] = f(r)$.

Now, suppose that $\lim_{x \to r}[f(x)] = f(r)$. Let $\epsilon > 0$. Then there is $\delta > 0$ such that $0 < |x - r| < \delta$ implies $|f(x) - f(r)| < \epsilon$. Let $x \in \mathbb{R}$ satisfy $|x - r| < \delta$. Then $0 < |x - r| < \delta$ or $x = r$. If $0 < |x - r| < \delta$, then $|f(x) - f(r)| < \epsilon$. If $x = r$, then $|f(x) - f(r)| = |f(r) - f(r)| = |0| = 0 < \epsilon$. Since $\epsilon > 0$ was arbitrary, f is continuous at r. □

5. Prove that every polynomial function $p: \mathbb{R} \to \mathbb{R}$ is continuous on $\mathbb{R}$.

Proof: Let $r \in \mathbb{R}$. We first show that for all $n \in \mathbb{N}$ with $n \geq 1$, $\lim_{x \to r}[x^k] = r^k$.

Base case ($n = 1$): Let $\epsilon > 0$ be given and let $\delta = \epsilon$. Then $0 < |x - r| < \delta$ implies $|x - r| < \delta = \epsilon$. Since $\epsilon > 0$ was arbitrary, $\lim_{x \to r}[x] = r$.

Inductive step: Let $k \in \mathbb{N}$ and assume that $\lim_{x \to r}[x^k] = r^k$. By Theorem 8.14, we have

$$\lim_{x \to r}[x^{k+1}] = \lim_{x \to r}[x^k \cdot x] = \lim_{x \to r}[x^k] \cdot \lim_{x \to r}[x] = r^k \cdot r = r^{k+1}.$$

By the Principle of Mathematical Induction, for all $k \in \mathbb{N}$ with $n \geq 1$, $\lim_{x \to r}[x^k] = r^k$.

Now, let $p: \mathbb{R} \to \mathbb{R}$ be a polynomial, say $p(x) = a_n x^n + a_{n-1} x^{n-1} + \cdots + a_1 x + a_0$. By Problem 2, $\lim_{x \to r}[a_0] = a_0$. By the last paragraph and Problem 3, $\lim_{x \to r}[a_k x^k] = a_k \lim_{x \to r}[x^k] = a_k r^k$. Finally, using Theorem 8.13, we have

$$\begin{aligned}\lim_{x \to r}[p(x)] &= \lim_{x \to r}[a_n x^n + a_{n-1} x^{n-1} + \cdots + a_1 x + a_0] \\ &= \lim_{x \to r}[a_n x^n] + \lim_{x \to r}[a_{n-1} x^{n-1}] + \cdots + \lim_{x \to r}[a_1 x] + \lim_{x \to r}[a_0] \\ &= a_n r^n + a_{n-1} r^{n-1} + \cdots + a_1 r + a_0 = p(r).\end{aligned}$$

By Problem 4, p is continuous at r. Since $r \in \mathbb{R}$ was arbitrary, p is continuous on $\mathbb{R}$. □

LEVEL 3

6. Let $g: \mathbb{R} \to \mathbb{R}$ be defined by $g(x) = 2x^2 - 3x + 7$.

   (i) Prove that $\lim_{x \to 1} g(x) = 6$.

   (ii) Prove that g is continuous on $\mathbb{R}$.

Proofs:

(i) Let $\epsilon > 0$ and let $\delta = \min\left\{1, \frac{\epsilon}{3}\right\}$. Suppose that $0 < |x - 1| < \delta$. Then we have $|x - 1| < 1$, so that $-1 < x - 1 < 1$. Adding 1, we get $0 < x < 2$. Multiplying by 2, we have $0 < 2x < 4$. Subtracting 1 gives us $-1 < 2x - 1 < 3$. So, $-3 < 2x - 1 < 3$, and therefore, $|2x - 1| < 3$. Now, we have

$$|(2x^2 - 3x + 7) - 6| = |2x^2 - 3x + 1| = |2x - 1||x - 1| < 3\delta \leq 3 \cdot \frac{\epsilon}{3} = \epsilon.$$

So, $\lim_{x \to 1} g(x) = 6$. □

(ii) Let $a \in \mathbb{R}$. We will show that f is continuous at a. Let $\epsilon > 0$ and let $\delta = \min\left\{1, \frac{\epsilon}{M}\right\}$, where $M = \max\{|4a - 8|, |4a - 4|\}$. Suppose that $|x - a| < \delta$. Then we have $|x - a| < 1$, so that $-1 < x - a < 1$. Adding $2a - 3$, we get $2a - 4 < x + a - 3 < 2a - 2$. Multiplying by 2 yields $4a - 8 < 2(x + a - 3) < 4a - 4$. Therefore, $-M < 2(x + a - 3) < M$, or equivalently, $|2(x + a - 3)| < M$. Now, we have

$$\begin{aligned}|(2x^2 - 3x + 7) - (2a^2 - 3a + 7)| &= |2(x^2 - a^2) - 3(x - a)| = |x - a||2(x + a - 3)| \\ &< \delta M \leq \frac{\epsilon}{M} \cdot M = \epsilon.\end{aligned}$$

So, g is continuous at a. Since $a \in \mathbb{R}$ was arbitrary, g is continuous on $\mathbb{R}$. □

7. Suppose that $f, g\colon \mathbb{R} \to \mathbb{R}$, $a \in \mathbb{R}$, f is continuous at a, and g is continuous at $f(a)$. Prove that $g \circ f$ is continuous at a.

Proof: Let $f, g\colon \mathbb{R} \to \mathbb{R}$, let $a \in \mathbb{R}$, and suppose that f is continuous at a and g is continuous at $f(a)$. Let $\epsilon > 0$. Since g is continuous at $f(a)$, there is $\delta_1 > 0$ such that $|y - f(a)| < \delta_1$ implies $|g(y) - g(f(a))| < \epsilon$. Since f is continuous at a, there is $\delta_2 > 0$ such that $|x - a| < \delta_2$ implies $|f(x) - f(a)| < \delta_1$. Now, suppose that $|x - a| < \delta_2$. Then $|f(x) - f(a)| < \delta_1$. It follows that $|g(f(x)) - g(f(a))| < \epsilon$. Since $\epsilon > 0$ was arbitrary, $g \circ f$ is continuous at a. □

LEVEL 4

8. Let $h\colon \mathbb{R} \to \mathbb{R}$ be defined by $h(x) = \frac{x^3 - 4}{x^2 + 1}$. Prove that $\lim_{x \to 2} h(x) = \frac{4}{5}$.

Proof: Let $\epsilon > 0$ and let $\delta = \min\left\{1, \frac{2\epsilon}{15}\right\}$. Suppose that $0 < |x - 2| < \delta$. Then we have $|x - 2| < 1$, so that $-1 < x - 2 < 1$. Adding 2, we get $1 < x < 3$. So, $23 < 5x^2 + 6x + 12 < 75$ and therefore, $-75 < 5x^2 + 6x + 12 < 75$. So, $|5x^2 + 6x + 12| < 75$. Also, $2 < x^2 + 1 < 10$. In particular, we have $x^2 + 1 > 2$, and so, $\frac{1}{x^2+1} < \frac{1}{2}$. Now, we have

$$\begin{aligned}\left|\frac{x^3 - 4}{x^2 + 1} - \frac{4}{5}\right| &= \left|\frac{5(x^3 - 4)}{5(x^2 + 1)} - \frac{4(x^2 + 1)}{5(x^2 + 1)}\right| = \left|\frac{5x^3 - 4x^2 - 24}{5(x^2 + 1)}\right| \\ &= \frac{|5x^2 + 6x + 12||x - 2|}{5(x^2 + 1)} < \frac{75\delta}{5 \cdot 2} \leq \frac{75}{10} \cdot \frac{2\epsilon}{15} = \epsilon.\end{aligned}$$

So, $\lim_{x\to 2} h(x) = \frac{4}{5}$. □

9. Let $k\colon (0,\infty) \to \mathbb{R}$ be defined by $k(x) = \sqrt{x}$.
   (i) Prove that $\lim_{x\to 25} k(x) = 5$.
   (ii) Prove that f is continuous on $(0,\infty)$.
   (iii) Is f uniformly continuous on $(0,\infty)$?

Proofs:

(i) Let $\epsilon > 0$ and let $\delta = \min\{1, (5+\sqrt{24})\epsilon\}$. Suppose that $0 < |x-25| < \delta$. Then we have $|x-25| < 1$, so that $-1 < x - 25 < 1$. Adding 25, we get $24 < x < 26$. Taking square roots, we have $\sqrt{24} < \sqrt{x} < \sqrt{26}$. Adding 5 gives us $5+\sqrt{24} < \sqrt{x}+5 < 5+\sqrt{26}$. So, $\frac{1}{\sqrt{x}+5} < \frac{1}{5+\sqrt{24}}$. Now, we have

$$\left|\sqrt{x}-5\right| = \left|\frac{(\sqrt{x}-5)(\sqrt{x}+5)}{\sqrt{x}+5}\right| = \frac{|x-25|}{\sqrt{x}+5} < \frac{\delta}{5+\sqrt{24}} \leq \frac{1}{5+\sqrt{24}} \cdot (5+\sqrt{24})\epsilon = \epsilon.$$

So, $\lim_{x\to 25} k(x) = 5$. □

(ii) Let $a \in \mathbb{R}$. We will show that f is continuous at a. Let $\epsilon > 0$, let $\delta = \min\{1, \epsilon\sqrt{a}\}$, and let $x \in (0,\infty)$ satisfy $|x-a| < \delta$. Then we have $|x-a| < 1$, so that $-1 < x - a < 1$. Adding a, we get $a-1 < x < a+1$. Since $x \in (0,\infty)$, we have $0 < x < a+1$. Taking square roots, we have $0 < \sqrt{x} < \sqrt{a+1}$. Adding $\sqrt{a}$ gives us $\sqrt{a} < \sqrt{x}+\sqrt{a} < \sqrt{a+1}+\sqrt{a}$. Therefore, $\frac{1}{\sqrt{x}+\sqrt{a}} < \frac{1}{\sqrt{a}}$. Now, we have

$$\left|\sqrt{x}-\sqrt{a}\right| = \left|\frac{(\sqrt{x}-\sqrt{a})(\sqrt{x}+\sqrt{a})}{\sqrt{x}+\sqrt{a}}\right| = \frac{|x-a|}{\sqrt{x}+\sqrt{a}} < \frac{\delta}{\sqrt{a}} \leq \frac{\epsilon\sqrt{a}}{\sqrt{a}} = \epsilon.$$

So, f is continuous at a. Since $a \in \mathbb{R}$ was arbitrary, f is continuous on $\mathbb{R}$. □

(iii) Let $\epsilon > 0$, let $\delta = \epsilon^2$, and let $x, y \in (0,\infty)$ satisfy $|x-y| < \delta$. Then we have

$$\left|\sqrt{x}-\sqrt{y}\right| = \sqrt{\left(\sqrt{x}-\sqrt{y}\right)^2} \leq \sqrt{\left|\sqrt{x}-\sqrt{y}\right|\left|\sqrt{x}+\sqrt{y}\right|} = \sqrt{|x-y|} < \sqrt{\delta} = \sqrt{\epsilon^2} = \epsilon.$$

So, f is uniformly continuous on $\mathbb{R}$. □

10. Let $f\colon \mathbb{R} \to \mathbb{R}$ be defined by $f(x) = x^2$. Prove that f is continuous on $\mathbb{R}$, but not uniformly continuous on $\mathbb{R}$.

Proof: Let $a \in \mathbb{R}$. We will show that f is continuous at a. Let $\epsilon > 0$ and let $\delta = \min\left\{1, \frac{\epsilon}{M}\right\}$, where $M = \max\{|2a-1|, |2a+1|\}$. Suppose that $|x-a| < \delta$. Then $|x-a| < 1$, so that $-1 < x - a < 1$. Adding $2a$, we get $2a-1 < x+a < 2a+1$. So, $-M < x+a < M$, or equivalently, $|x+a| < M$. Now, we have

$$|x^2 - a^2| = |x - a||x + a| < \delta \cdot M \leq \frac{\epsilon}{M} \cdot M = \epsilon.$$

So, f is continuous at a. Since $a \in \mathbb{R}$ was arbitrary, f is continuous on $\mathbb{R}$.

To see that f is not uniformly continuous, let $\epsilon = 1$ and let $\delta > 0$. Let $x = \frac{1}{\delta}$ and $y = \frac{1}{\delta} + \frac{\delta}{2}$. Then we have $|x - y| = \left|\frac{1}{\delta} - \left(\frac{1}{\delta} + \frac{\delta}{2}\right)\right| = \frac{\delta}{2}$, but

$$|f(x) - f(y)| = |x^2 - y^2| = \left|\frac{1}{\delta^2} - \left(\frac{1}{\delta} + \frac{\delta}{2}\right)^2\right| = \left|\frac{1}{\delta^2} - \frac{1}{\delta^2} - 1 - \frac{\delta^2}{4}\right| = 1 + \frac{\delta^2}{4} > 1 = \epsilon.$$

So, f is **not** uniformly continuous on $\mathbb{R}$ (and in fact, not uniformly continuous on $(0, \infty)$ since we only needed positive values of x and y to violate the definition of uniform continuity). □

11. Let $A \subseteq \mathbb{R}$, let $f: A \to \mathbb{R}$, let $r \in \mathbb{R}$, and suppose that $\lim\limits_{x \to r}[f(x)] > 0$. Prove that there is a deleted neighborhood N of r such that $f(x) > 0$ for all $x \in N$.

Proof: Suppose that $\lim\limits_{x \to r}[f(x)] = L$ with $L > 0$. Let $\epsilon = \frac{L}{2}$. There is $\delta > 0$ such that $0 < |x - r| < \delta$ implies $|f(x) - L| < \epsilon$. Consider $N_\delta^{\odot}(r) = (r - \delta, r) \cup (r, r + \delta)$. Let $x \in N_\delta^{\odot}(r)$. Then we have $x \in (r - \delta, r) \cup (r, r + \delta)$, so that $0 < |x - r| < \delta$. It follows that $|f(x) - L| < \epsilon = \frac{L}{2}$. So, we have $-\frac{L}{2} < f(x) - L < \frac{L}{2}$, or equivalently, $L - \frac{L}{2} < f(x) < L + \frac{L}{2}$. Since $L - \frac{L}{2} = \frac{L}{2}$ and $L + \frac{L}{2} = \frac{3L}{2}$, we have $\frac{L}{2} < f(x) < \frac{3L}{2}$. In particular, we have $f(x) > \frac{L}{2} > 0$. Since $x \in N_\delta^{\odot}(r)$ was arbitrary, we have shown that for all $x \in N_\delta^{\odot}(r)$, $f(x) > 0$. □

12. Let $A \subseteq \mathbb{R}$, let $f: A \to \mathbb{R}$, let $r \in \mathbb{R}$, and suppose that $\lim\limits_{x \to r}[f(x)]$ is a finite real number. Prove that there is $M \in \mathbb{R}$ and an open interval (a, b) containing r such that $|f(x)| \leq M$ for all $x \in (a, b) \setminus \{r\}$.

Proof: Let $A \subseteq \mathbb{R}$, $f: A \to \mathbb{R}$, $r \in \mathbb{R}$, $\lim\limits_{x \to r}[f(x)] = L$, and let $\epsilon = 1$. Then there is $\delta > 0$ such that $0 < |x - r| < \delta$ implies $|f(x) - L| < 1$, or $-1 < f(x) - L < 1$, or $L - 1 < f(x) < L + 1$. Let $a = r - \delta$, $b = r + \delta$, and $M = \max\{|L - 1|, |L + 1\}$. If $x \in (a, b) \setminus \{r\}$, then $r - \delta < x < r + \delta$ and $x \neq r$. So, $0 < |x - r| < \delta$. Therefore, $L - 1 < f(x) < L + 1$. Since $M \geq |L - 1| \geq 1 - L$, we have $-M \leq L - 1$. Also, $M \geq |L + 1| \geq L + 1$. So, we have $-M < f(x) < M$, or equivalently, $|f(x)| < M$. Since $x \in (a, b) \setminus \{r\}$ was arbitrary, $|f(x)| < M$ for all $x \in (a, b) \setminus \{r\}$. □

13. Let $A \subseteq \mathbb{R}$, let $f, g, h: A \to \mathbb{R}$, let $r \in \mathbb{R}$, let $f(x) \leq g(x) \leq h(x)$ for all $x \in A \setminus \{r\}$, and suppose that $\lim\limits_{x \to r}[f(x)] = \lim\limits_{x \to r}[h(x)] = L$. Prove that $\lim\limits_{x \to r}[g(x)] = L$.

Proof: Let $\epsilon > 0$. Since $\lim_{x \to r}[f(x)] = L$, there is $\delta_1 > 0$ such that $0 < |x - r| < \delta_1$ implies $|f(x) - L| < \epsilon$. Since $\lim_{x \to r}[h(x)] = L$, there is $\delta_2 > 0$ such that $0 < |x - r| < \delta_2$ implies $|h(x) - L| < \epsilon$. Let $\delta = \min\{\delta_1, \delta_2\}$ and let $0 < |x - r| < \delta$. Then $0 < |x - r| < \delta_1$, so that $|f(x) - L| < \epsilon$, or equivalently, $-\epsilon < f(x) - L < \epsilon$, or $L - \epsilon < f(x) < L + \epsilon$. We will need only that $L - \epsilon < f(x)$. Similarly, we have $0 < |x - r| < \delta_2$, so that $|h(x) - L| < \epsilon$, or equivalently, $-\epsilon < h(x) - L < \epsilon$, or $L - \epsilon < h(x) < L + \epsilon$. We will need only that $h(x) < L + \epsilon$. Now, we have $L - \epsilon < f(x) \leq g(x) \leq h(x) < L + \epsilon$. So, $-\epsilon < g(x) - L < \epsilon$, or equivalently, $|g(x) - L| < \epsilon$. Since $\epsilon > 0$ was arbitrary, $\lim_{x \to r}[g(x)] = L$. □

LEVEL 5

14. Let $A \subseteq \mathbb{R}$, let $f, g: A \to \mathbb{R}$ such that $g(x) \neq 0$ for all $x \in A$, let $r \in \mathbb{R}$, and suppose that $\lim_{x \to r}[f(x)]$ and $\lim_{x \to r}[g(x)]$ are both finite real numbers such that $\lim_{x \to r}[g(x)] \neq 0$. Prove that $\lim_{x \to r}\left[\frac{f(x)}{g(x)}\right] = \frac{\lim_{x \to r} f(x)}{\lim_{x \to r} g(x)}$.

Proof: Suppose that $\lim_{x \to r}[f(x)] = L$ and $\lim_{x \to r}[g(x)] = K$, and let $\epsilon > 0$. Since $\lim_{x \to r}[g(x)] = K$, there is $\delta_1 > 0$ such that $0 < |x - r| < \delta_1$ implies $|g(x) - K| < \frac{|K|}{2}$. Now, $|g(x) - K| < \frac{|K|}{2}$ is equivalent to $-\frac{|K|}{2} < g(x) - K < \frac{|K|}{2}$, or by adding K, $K - \frac{|K|}{2} < g(x) < K + \frac{|K|}{2}$. If $K > 0$, we have $\frac{K}{2} < g(x) < \frac{3K}{2}$. If $K < 0$, we have $\frac{3K}{2} < g(x) < \frac{K}{2}$. In both cases, we have $\frac{|K|}{2} < |g(x)| < \frac{3|K|}{2}$. Let $M = \frac{|K|}{2}$. Then $|g(x)| > M$, and so, $\frac{1}{|g(x)|} < \frac{1}{M}$.

Now, since $\lim_{x \to r}[f(x)] = L$, there is $\delta_2 > 0$ such that $0 < |x - r| < \delta_2$ implies $|f(x) - L| < \frac{M|K|\epsilon}{|K|+|L|}$. Since $\lim_{x \to r}[g(x)] = K$, there is $\delta_3 > 0$ such that $0 < |x - r| < \delta_3$ implies $|g(x) - K| < \frac{M|K|\epsilon}{|K|+|L|}$. Let $\delta = \min\{\delta_1, \delta_2, \delta_3\}$ and suppose that $0 < |x - r| < \delta$. Then since $\delta \leq \delta_1$, $\frac{1}{|g(x)|} < \frac{1}{M}$. Since $\delta \leq \delta_2$,

$|f(x) - L| < \frac{M|K|\epsilon}{|K|+|L|}$. Since $\delta \leq \delta_3$, $|g(x) - K| < \frac{M|K|\epsilon}{|K|+|L|}$. By the Triangle Inequality (and SACT), we have

$$\left|\frac{f(x)}{g(x)} - \frac{L}{K}\right| = \left|\frac{Kf(x) - Lg(x)}{Kg(x)}\right| = \left|\frac{Kf(x) - KL + KL - Lg(x)}{Kg(x)}\right| = \left|\frac{Kf(x) - KL}{Kg(x)} + \frac{KL - Lg(x)}{Kg(x)}\right|$$

$$\leq \left|\frac{Kf(x) - KL}{Kg(x)}\right| + \left|\frac{KL - Lg(x)}{Kg(x)}\right| = \left|\frac{f(x) - L}{g(x)}\right| + \left|\frac{L}{K}\right|\left|\frac{K - g(x)}{g(x)}\right| = \left|\frac{f(x) - L}{g(x)}\right| + \left|\frac{L}{K}\right|\left|\frac{g(x) - K}{g(x)}\right|$$

$$= \frac{1}{|g(x)|}\left(|f(x) - L| + \left|\frac{L}{K}\right||g(x) - K|\right) < \frac{1}{M}\left(\frac{M|K|\epsilon}{|K| + |L|} + \left|\frac{L}{K}\right|\frac{M|K|\epsilon}{|K| + |L|}\right) = \frac{1}{M} \cdot \frac{M|K|\epsilon}{|K| + |L|}\left(1 + \left|\frac{L}{K}\right|\right)$$

$$= \frac{|K|\epsilon}{|K| + |L|}\left(\frac{|K| + |L|}{|K|}\right) = \epsilon.$$

So, $\lim_{x \to r}\left[\frac{f(x)}{g(x)}\right] = \frac{L}{K} = \frac{\lim_{x \to r}[f(x)]}{\lim_{x \to r}[g(x)]}$. □

15. Give a reasonable equivalent definition for each of the following limits (like what was done in Theorem 8.20). r and L are finite real numbers.

(i) $\lim_{x \to r} f(x) = -\infty$

(ii) $\lim_{x \to +\infty} f(x) = L$

(iii) $\lim_{x \to -\infty} f(x) = L$

(iv) $\lim_{x \to +\infty} f(x) = +\infty$

(v) $\lim_{x \to +\infty} f(x) = -\infty$

(vi) $\lim_{x \to -\infty} f(x) = +\infty$

(vii) $\lim_{x \to -\infty} f(x) = -\infty$

Equivalent definitions:

(i) $\lim_{x \to r} f(x) = -\infty$ if and only if $\forall M > 0\ \exists \delta > 0\ (0 < |x - r| < \delta \to f(x) < -M)$.

(ii) $\lim_{x \to +\infty} f(x) = L$ if and only if $\forall \epsilon > 0\ \exists K > 0\ (x > K \to |f(x) - L| < \epsilon)$.

(iii) $\lim_{x \to -\infty} f(x) = L$ if and only if $\forall \epsilon > 0\ \exists K > 0\ (x < -K \to |f(x) - L| < \epsilon)$.

(iv) $\lim_{x \to +\infty} f(x) = +\infty$ if and only if $\forall M > 0\ \exists K > 0\ (x > K \to f(x) > M)$.

(v) $\lim_{x \to +\infty} f(x) = -\infty$ if and only if $\forall M > 0\ \exists K > 0\ (x > K \to f(x) < -M)$.

(vi) $\lim_{x \to -\infty} f(x) = +\infty$ if and only if $\forall M > 0\ \exists K > 0\ (x < -K \to f(x) > M)$.

(vii) $\lim_{x \to -\infty} f(x) = -\infty$ if and only if $\forall M > 0\ \exists K > 0\ (x < -K \to f(x) < -M)$.

16. Let $f(x) = -x^2 + x + 1$. Use the $M - K$ definition of an infinite limit (that you came up with in Problem 15) to prove $\lim_{x \to +\infty} f(x) = -\infty$.

Proof: Let $M > 0$ and let $K = \max\left\{\frac{1}{2}, \frac{1}{2} + \sqrt{M + \frac{5}{4}}\right\}$. Suppose that $x > K$. Then $x - \frac{1}{2} > \sqrt{M + \frac{5}{4}}$, and so, $\left(x - \frac{1}{2}\right)^2 > M + \frac{5}{4}$. So, $x^2 - x + \frac{1}{4} > M + \frac{5}{4}$. Thus, $x^2 - x - 1 > M$. Therefore, $-x^2 + x + 1 < -M$. That is, $f(x) < -M$. So, $\lim_{x \to +\infty} g(x) = -\infty$. □

17. Give a reasonable definition for each of the following limits (like what was done in Theorem 8.22). r and L are finite real numbers.

(i) $\lim_{x \to r^-} f(x) = L$

(ii) $\lim_{x \to r^+} f(x) = +\infty$

(iii) $\lim_{x \to r^+} f(x) = -\infty$

(iv) $\lim_{x \to r^-} f(x) = +\infty$

(v) $\lim_{x \to r^-} f(x) = -\infty$

Definitions:

(i) $\lim_{x \to r^-} f(x) = L$ if and only if $\forall \epsilon > 0\, \exists \delta > 0\, (-\delta < x - r < 0 \to |f(x) - L| < \epsilon)$.

(ii) $\lim_{x \to r^+} f(x) = +\infty$ if and only if $\forall M > 0\, \exists \delta > 0\, (0 < x - r < \delta \to f(x) > M)$.

(iii) $\lim_{x \to r^+} f(x) = -\infty$ if and only if $\forall M > 0\, \exists \delta > 0\, (0 < x - r < \delta \to f(x) < -M)$.

(iv) $\lim_{x \to r^-} f(x) = +\infty$ if and only if $\forall M > 0\, \exists \delta > 0\, (-\delta < x - r < 0 \to f(x) > M)$.

(v) $\lim_{x \to r^-} f(x) = -\infty$ if and only if $\forall M > 0\, \exists \delta > 0\, (-\delta < x - r < 0 \to f(x) < -M)$.

18. Use the $M - \delta$ definition of a one-sided limit (that you came up with in Problem 17) to prove that $\lim_{x \to 3^-} \frac{1}{x-3} = -\infty$.

Proof: Let $M > 0$ and let $\delta = \frac{1}{M}$. If $-\delta < x - 3 < 0$, then $-\frac{1}{M} < x - 3 < 0$, and so, we have $\frac{1}{x-3} < -M$. Since $M > 0$ was arbitrary, $\lim_{x \to 3^-} \frac{1}{x-3} = -\infty$. □

19. Let $f(x) = \frac{x+1}{(x-1)^2}$. Prove that

(i) $\lim_{x \to +\infty} f(x) = 0$.

(ii) $\lim_{x \to 1^+} f(x) = +\infty$.

Proofs:

(i) Let $\epsilon > 0$ and let $K = \max\left\{2, 1 + \frac{3}{\epsilon}\right\}$. Let $x > K$. Then $x - 1 > 1 + \frac{3}{\epsilon} - 1 = \frac{3}{\epsilon}$, and therefore, $\frac{1}{x-1} < \frac{\epsilon}{3}$. Also, since $x > 2$, $(x-1)^2 - (x-1) = (x-1)(x-1-1) = (x-1)(x-2) > 0$ (because $x - 1 > 2 - 1 = 1 > 0$ and $x - 2 > 2 - 2 > 0$). Thus, $(x-1)^2 > x - 1$, and so, $\frac{1}{(x-1)^2} < \frac{1}{x-1} < \frac{\epsilon}{3}$. It follows from the triangle inequality (and SACT) that

(ii)

$$\left|\frac{x+1}{(x-1)^2}-0\right|=\left|\frac{x-1+2}{(x-1)^2}\right|=\left|\frac{x-1}{(x-1)^2}+\frac{2}{(x-1)^2}\right|=\left|\frac{1}{x-1}+\frac{2}{(x-1)^2}\right|$$

$$\leq\left|\frac{1}{x-1}\right|+\left|\frac{2}{(x-1)^2}\right|=\frac{1}{x-1}+2\frac{1}{(x-1)^2}<\frac{\epsilon}{3}+2\cdot\frac{\epsilon}{3}=3\cdot\frac{\epsilon}{3}=\epsilon.$$

So, $\lim_{x\to+\infty} f(x)=0$. □

(iii) Let $M>0$ and let $\delta=\min\left\{1,\frac{3}{M}\right\}$. If $0<x-1<\delta$, then $0<x-1<\frac{3}{M}$, and so, we have $\frac{1}{x-1}>\frac{M}{3}$. Since $0<x-1<1$, $(x-1)^2<x-1$, and so, $\frac{1}{(x-1)^2}>\frac{1}{x-1}$. So, we have

$$\frac{x+1}{(x-1)^2}=\frac{x-1+2}{(x-1)^2}=\frac{x-1}{(x-1)^2}+\frac{2}{(x-1)^2}=\frac{1}{x-1}+\frac{2}{(x-1)^2}$$

$$>\frac{1}{x-1}+\frac{2}{x-1}=\frac{3}{x-1}>3\cdot\frac{M}{3}=M.$$

So, $\lim_{x\to 1} f(x)=+\infty$. □

20. Let $f:\mathbb{R}\to\mathbb{R}$ be defined by $f(x)=\begin{cases}0 & \text{if } x \text{ is rational}\\ 1 & \text{if } x \text{ is irrational}\end{cases}$. Prove that for all $r\in\mathbb{R}$, $\lim_{x\to r}[f(x)]$ does not exist.

Proof: Let $r\in\mathbb{R}$, let $\epsilon=\frac{1}{2}$, and let $\delta>0$. By the Density Theorem (Theorem 6.17) and Problem 19 from Problem Set 6, there is a rational number x and an irrational number y such that $r<x,y<r+\delta$. So, we have $0<|x-r|<\delta$ and $0<|y-r|<\delta$. We also have $f(x)=0$ and $f(y)=1$. Let $L\in\mathbb{R}$. If $\lim_{x\to r}[f(x)]=L$, then $|f(x)-L|<\frac{1}{2}$ and $|f(y)-L|<\frac{1}{2}$. But then we would have $|f(x)-f(y)|=|f(x)-L+L-f(y)|\leq|f(x)-L|+|L-f(y)|<\frac{1}{2}+\frac{1}{2}=1$. However, $|f(x)-f(y)|=|1-0|=1$. Since $1<1$ is false, $\lim_{x\to r}[f(x)]$ does not equal L. Since $L\in\mathbb{R}$ was arbitrary, $\lim_{x\to r}[f(x)]$ does not exist. □

21. Let $A\subseteq\mathbb{C}$, let $f:A\to\mathbb{C}$, let $L=j+ki\in\mathbb{C}$, and let $a=b+ci\in\mathbb{C}$ be a point such that A contains some deleted neighborhood of a. Suppose that $f(x+yi)=u(x,y)+iv(x,y)$. Prove that $\lim_{z\to a} f(z)=L$ if and only if $\lim_{(x,y)\to(b,c)} u(x,y)=j$ and $\lim_{(x,y)\to(b,c)} v(x,y)=k$.

Proof: Suppose that $\lim_{z\to a} f(z)=L$ and let $\epsilon>0$. Then there is $\delta>0$ such that $0<|z-a|<\delta$ implies $|f(z)-L|<\epsilon$. Now,

$$|z-a|=|(x+yi)-(b+ci)|=|(x-b)+(y-c)i|=\sqrt{(x-b)^2+(y-c)^2}$$

Also, $|f(z)-L|=\left|\big(u(x,y)+iv(x,y)\big)-(j+ki)\right|=|(u(x,y)-j)+(v(x,y)-k)i|$.

So, if $0 < \sqrt{(x-b)^2 + (y-c)^2} < \delta$, then $0 < |z - a| < \delta$, and therefore, $|f(z) - L| < \epsilon$. It follows that

$$|u(x,y) - j| \leq |(u(x,y) - j) + (v(x,y) - k)i| = |f(z) - L| < \epsilon$$

and

$$|v(x,y) - k| \leq |(u(x,y) - j) + (v(x,y) - k)i| = |f(z) - L| < \epsilon$$

Therefore, $\lim\limits_{(x,y)\to(b,c)} u(x,y) = j$ and $\lim\limits_{(x,y)\to(b,c)} v(x,y) = k$.

Conversely, suppose that $\lim\limits_{(x,y)\to(b,c)} u(x,y) = j$ and $\lim\limits_{(x,y)\to(b,c)} v(x,y) = k$ and let $\epsilon > 0$. Then there are $\delta_1, \delta_2 > 0$ such that

$$0 < \sqrt{(x-b)^2 + (y-c)^2} < \delta_1 \text{ implies } |u(x,y) - j| < \tfrac{\epsilon}{2}$$

and

$$0 < \sqrt{(x-b)^2 + (y-c)^2} < \delta_2 \text{ implies } |v(x,y) - k| < \tfrac{\epsilon}{2}$$

Let $\delta = \min\{\delta_1, \delta_2\}$ and assume that $0 < |z - a| < \delta$. Since $|z - a| = \sqrt{(x-b)^2 + (y-c)^2}$ and $\delta \leq \delta_1, \delta_2$, we have $|u(x,y) - j| < \frac{\epsilon}{2}$ and $|v(x,y) - k| < \frac{\epsilon}{2}$. It follows that

$$|f(z) - L| = \big|\big(u(x,y) + iv(x,y)\big) - (j + ki)\big| = |(u(x,y) - j) + (v(x,y) - k)i|$$
$$\leq |u(x,y) - j| + |v(x,y) - k| < \frac{\epsilon}{2} + \frac{\epsilon}{2} = \epsilon.$$

Therefore, $\lim\limits_{z\to a} f(z) = L$. □

22. Give a reasonable definition for each of the following limits (like what was done right before Theorem 8.28). L is a finite real number.

(i) $\lim\limits_{z\to\infty} f(z) = L$

(ii) $\lim\limits_{z\to\infty} f(z) = \infty$

Equivalent definitions:

(i) $\lim\limits_{z\to\infty} f(z) = L$ if and only if $\forall\epsilon > 0\ \exists\delta > 0\ \left(|z| > \frac{1}{\delta} \to |f(z) - L| < \epsilon\right)$.

(ii) $\lim\limits_{z\to\infty} f(z) = \infty$ if and only if $\forall\epsilon > 0\ \exists\delta > 0\ \left(|z| > \frac{1}{\delta} \to |f(z)| > \frac{1}{\epsilon}\right)$.

23. Prove each of the following:

(i) $\lim\limits_{z\to\infty} f(z) = L$ if and only $\lim\limits_{z\to 0} f\left(\frac{1}{z}\right) = L$

(ii) $\lim\limits_{z\to\infty} f(z) = \infty$ if and only $\lim\limits_{z\to 0} \frac{1}{f\left(\frac{1}{z}\right)} = 0$.

Proofs:

(i) Suppose that $\lim_{z\to\infty} f(z) = L$ and let $\epsilon > 0$. There is $\delta > 0$ so that $|z| > \frac{1}{\delta} \to |f(z) - L| < \epsilon$. If we let $w = \frac{1}{z}$, we have $z = \frac{1}{w}$, and therefore, $\left|\frac{1}{w}\right| > \frac{1}{\delta} \to \left|f\left(\frac{1}{w}\right) - L\right| < \epsilon$. But $\left|\frac{1}{w}\right| > \frac{1}{\delta}$ is equivalent to $0 < |w| < \delta$. So, $0 < |w - 0| < \delta \to \left|f\left(\frac{1}{w}\right) - L\right| < \epsilon$. Thus, $\lim_{w\to 0} f\left(\frac{1}{w}\right) = L$. This is equivalent to $\lim_{z\to 0} f\left(\frac{1}{z}\right) = L$.

Conversely, suppose that $\lim_{z\to 0} f\left(\frac{1}{z}\right) = L$ and let $\epsilon > 0$. Then there is $\delta > 0$ so that $0 < |z - 0| < \delta \to \left|f\left(\frac{1}{z}\right) - L\right| < \epsilon$. If we let $w = \frac{1}{z}$, then $z = \frac{1}{w}$, and therefore, we have $0 < \left|\frac{1}{w}\right| < \delta \to |f(w) - L| < \epsilon$. Now, $0 < \left|\frac{1}{w}\right| < \delta$ is equivalent to $|w| > \frac{1}{\delta}$. So, we have $|w| > \frac{1}{\delta} \to |f(w) - L| < \epsilon$. Therefore, $\lim_{w\to\infty} f(w) = L$, or equivalently, $\lim_{z\to\infty} f(z) = L$. □

(ii) Suppose that $\lim_{z\to\infty} f(z) = \infty$ and let $\epsilon > 0$. There is $\delta > 0$ so that $|z| > \frac{1}{\delta} \to |f(z)| > \frac{1}{\epsilon}$. If we let $w = \frac{1}{z}$, we have $z = \frac{1}{w}$, and therefore, $\left|\frac{1}{w}\right| > \frac{1}{\delta} \to \left|f\left(\frac{1}{w}\right)\right| > \frac{1}{\epsilon}$. But, $\left|\frac{1}{w}\right| > \frac{1}{\delta}$ is equivalent to $0 < |w| < \delta$ and $\left|f\left(\frac{1}{w}\right)\right| > \frac{1}{\epsilon}$ is equivalent to $\left|\frac{1}{f\left(\frac{1}{w}\right)}\right| < \epsilon$. So, $0 < |w| < \delta \to \left|\frac{1}{f\left(\frac{1}{w}\right)}\right| < \epsilon$, and therefore, $\lim_{w\to 0} \frac{1}{f\left(\frac{1}{w}\right)} = 0$. This is equivalent to $\lim_{z\to 0} \frac{1}{f\left(\frac{1}{z}\right)} = 0$.

Now, let $\lim_{z\to 0} \frac{1}{f\left(\frac{1}{z}\right)} = 0$ and let $\epsilon > 0$. There is $\delta > 0$ so that $0 < |z - 0| < \delta \to \left|\frac{1}{f\left(\frac{1}{z}\right)} - 0\right| < \epsilon$. If we let $w = \frac{1}{z}$, we have $z = \frac{1}{w}$, and therefore, $0 < |z - 0| < \delta$ is equivalent to $0 < \left|\frac{1}{w}\right| < \delta$, or equivalently, $|w| > \frac{1}{\delta}$. Also, $\left|\frac{1}{f\left(\frac{1}{z}\right)} - 0\right| < \epsilon$ is equivalent to $\left|\frac{1}{f(w)}\right| < \epsilon$, which in turn is equivalent to $|f(w)| > \frac{1}{\epsilon}$. So, $|w| > \frac{1}{\delta} \to |f(w)| > \frac{1}{\epsilon}$. Thus, $\lim_{w\to\infty} f(w) = \infty$. This is equivalent to $\lim_{z\to\infty} f(z) = \infty$. □

24. Let $f\colon \mathbb{R}^2 \to \mathbb{R}$ be defined by $f(x, y) = xy$. Prove that f is continuous. Then generalize this result to the function $f\colon \mathbb{R}^n \to \mathbb{R}$ defined by $f(\mathbf{x}) = x_1 \cdot x_2 \cdots x_n$.

Proof: Let $(a, b) \in \mathbb{R}^2$. We will show that f is continuous at (a, b). Let $\epsilon > 0$ and let $\delta = \min\left\{1, \frac{\epsilon}{1+|a|+|b|}\right\}$. Suppose that

$$|(x, y) - (a, b)| < \delta, \text{ or equivalently, } \sqrt{(x-a)^2 + (y-b)^2} < \delta.$$

Then $|x| = |(x - a) + a| \leq |x - a| + |a| \leq \sqrt{(x-a)^2 + (y-b)^2} + |a| < \delta + |a| \leq 1 + |a|$. So,

$$|xy - ab| = |xy - xb + xb - ab| \leq |xy - xb| + |xb - ab| = |x||y - b| + |b||x - a|$$
$$\leq (1 + |a|)\sqrt{(x-a)^2 + (y-b)^2} + |b|\sqrt{(x-a)^2 + (y-b)^2} < (1 + |a| + |b|)\delta$$
$$\leq (1 + |a| + |b|) \cdot \frac{\epsilon}{(1 + |a| + |b|)} = \epsilon.$$

So, f is continuous at (a,b). Since $(a,b) \in \mathbb{R}^2$ was arbitrary, f is continuous.

We now prove that if $h_1, h_2, \dots, h_m: \mathbb{R}^n \to \mathbb{R}$ are continuous and $g: \mathbb{R}^n \to \mathbb{R}^m$ is defined by $g(x_1, x_2, \dots, x_n) = \big(h_1(x_1, \dots, x_n), h_2(x_1, \dots, x_n), \dots, h_m(x_1, \dots, x_n)\big)$, then g is continuous. To see this, let $(a_1, \dots, a_n) \in \mathbb{R}^n$ and let $\epsilon > 0$. Since each h_j is continuous, for each $j = 1, 2, \dots, m$, we can find $\delta_j > 0$ such that

$$0 < \sqrt{(x_1 - a_1)^2 + \cdots + (x_n - a_n)^2} < \delta_j \text{ implies } \left|h_j(x_1, \dots, x_n) - h_j(a_1, \dots, a_n)\right| < \frac{\epsilon}{\sqrt{m}}.$$

Let $\delta = \min\{\delta_1, \dots, \delta_m\}$. Suppose that $0 < \sqrt{(x_1 - a_1)^2 +\cdot +(x_n - a_n)^2} < \delta$. Then we have

$$\sqrt{\big(h_1(x_1, \dots, x_n) - h_1(a_1, \dots, a_n)\big)^2 + \cdots + \big(h_m(x_1, \dots, x_n) - h_m(a_1, \dots, a_n)\big)^2}$$

$$< \sqrt{\left(\frac{\epsilon}{\sqrt{m}}\right)^2 + \cdots + \left(\frac{\epsilon}{\sqrt{m}}\right)^2} = \sqrt{m \cdot \frac{\epsilon^2}{m}} = \sqrt{\epsilon^2} = \epsilon.$$

Therefore, g is continuous at $(a_1, \dots, a_n)$. Since $(a_1, \dots, a_n) \in \mathbb{R}^n$ was arbitrary, g is continuous.

We now prove that if $h: \mathbb{R}^m \to \mathbb{R}$ and $g: \mathbb{R}^n \to \mathbb{R}^m$ are continuous, then $h \circ g: \mathbb{R}^n \to \mathbb{R}$ is continuous. To see this, let $(a_1, \dots, a_n) \in \mathbb{R}^n$ and let $\epsilon > 0$. Since h is continuous at $g(a_1, \dots, a_n)$, there is $\delta_1 > 0$ such that $|(y_1, \dots, y_m) - g(a_1, \dots, a_n)| < \delta_1$ implies $\left|h(y_1, \dots, y_m) - h\big(g(a_1, \dots, a_n)\big)\right| < \epsilon$. Since g is continuous at $(a_1, \dots, a_n)$, there is $\delta_2 > 0$ such that $|(x_1, \dots, x_n) - (a_1, \dots, a_n)| < \delta_2$ implies $|g(x_1, \dots, x_n) - g(a_1, \dots, a_n)| < \delta_1$. Now, suppose that $|(x_1, \dots, x_n) - (a_1, \dots, a_n)| < \delta_2$. Then $|g(x_1, \dots, x_n) - g(a_1, \dots, a_n)| < \delta_1$. It follows that $\left|h\big(g(x_1, \dots, x_m)\big) - h\big(g(a_1, \dots, a_n)\big)\right| < \epsilon$. Since $\epsilon > 0$ was arbitrary, $h \circ g$ is continuous at $(a_1, \dots, a_n)$. Since $(a_1, \dots, a_n) \in \mathbb{R}^n$ was arbitrary, $h \circ g$ is continuous.

We now proceed by induction on $n \in \mathbb{N}$, with the base case of $n = 2$ having been proved above.

Assume that $f_k: \mathbb{R}^k \to \mathbb{R}$ is continuous, where f_k is defined by $f_k(\mathbf{x}) = x_1 \cdot x_2 \cdots x_k$, and let $f_{k+1}: \mathbb{R}^{k+1} \to \mathbb{R}$ be defined by $f_{k+1}(\mathbf{x}) = x_1 \cdot x_2 \cdots x_k \cdot x_{k+1}$. Let $g_k: \mathbb{R}^{k+1} \to \mathbb{R}^k$ be defined by $g_k(\mathbf{x}) = (x_1, x_2, \dots, x_{k-1}, x_k x_{k+1})$. Then we have

$$(f_k \circ g_k)(\mathbf{x}) = f_k\big(g_k(\mathbf{x})\big) = f_k(x_1, x_2, \dots, x_{k-1}, x_k x_{k+1}) = x_1 \cdot x_2 \cdots x_k \cdot x_{k+1} = f_{k+1}(\mathbf{x})$$

So, $f_{k+1} = f_k \circ g_k$. Therefore, the continuity of f_{k+1} will follow from the continuity of f_k and g_k.

Our inductive assumption was that f_k is continuous.

Now, observe that $g_k(x_1, x_2, \dots, x_{k+1}) = \big(h_1(x_1, \dots, x_{k+1}), h_2(x_1, \dots, x_{k+1}), \dots, h_k(x_1, \dots, x_{k+1})\big)$, where $h_j(x_1, \dots, x_{k+1}) = x_j$ for $j = 1, 2, \dots, k-1$ and $h_k(x_1, \dots, x_{k+1}) = x_k \cdot x_{k+1}$. The functions h_j for $j = 1, 2, \dots, k-1$ are continuous because $|x_j - a_j| \le \sqrt{(x_1 - a_1)^2 + \cdots (x_{k+1} - a_{k+1})^2}$. Finally, $h_k = b_k \circ c_k$, where $c_k: \mathbb{R}^{k+1} \to \mathbb{R}^2$ is defined by $c_k(x_1, \dots, x_{k+1}) = (x_k, x_{k+1})$ and $b_k: \mathbb{R}^2 \to \mathbb{R}$ is defined by $b_k(x,y) = xy$. The function c_k is continuous because

$$\sqrt{(x_k - a_k)^2 + \cdots + (x_{k+1} - a_{k+1})^2} \le \sqrt{(x_1 - a_1)^2 + \cdots + (x_{k+1} - a_{k+1})^2}$$

The continuity of b_k was the base case for the induction. □

Problem Set 9

LEVEL 1

1. Let $\mathcal{T}$ and $\mathcal{U}$ be topologies on a set S. Prove that $\mathcal{T} \cap \mathcal{U}$ is a topology on S.

Proof: Since $\mathcal{T}$ is a topology on S, $\emptyset \in \mathcal{T}$. Since $\mathcal{U}$ is a topology on S, $\emptyset \in \mathcal{U}$. Since $\emptyset \in \mathcal{T}$ and $\emptyset \in \mathcal{U}$, we have $\emptyset \in \mathcal{T} \cap \mathcal{U}$.

Since $\mathcal{T}$ is a topology on S, $S \in \mathcal{T}$. Since $\mathcal{U}$ is a topology on S, $S \in \mathcal{U}$. Since $S \in \mathcal{T}$ and $S \in \mathcal{U}$, we have $S \in \mathcal{T} \cap \mathcal{U}$.

Let $\boldsymbol{X} \subseteq \mathcal{T} \cap \mathcal{U}$. Then $\boldsymbol{X} \subseteq \mathcal{T}$ and $\boldsymbol{X} \subseteq \mathcal{U}$. Since $\boldsymbol{X} \subseteq \mathcal{T}$ and $\mathcal{T}$ is a topology on S, $\bigcup \boldsymbol{X} \in \mathcal{T}$. Since $\boldsymbol{X} \subseteq \mathcal{U}$ and $\mathcal{U}$ is a topology on S, $\bigcup \boldsymbol{X} \in \mathcal{U}$. Since $\bigcup \boldsymbol{X} \in \mathcal{T}$ and $\bigcup \boldsymbol{X} \in \mathcal{U}$, we have $\bigcup \boldsymbol{X} \in \mathcal{T} \cap \mathcal{U}$.

Let $\boldsymbol{Y} \subseteq \mathcal{T} \cap \mathcal{U}$ with $\boldsymbol{Y}$ finite. Then $\boldsymbol{Y} \subseteq \mathcal{T}$ and $\boldsymbol{Y} \subseteq \mathcal{U}$ Since $\boldsymbol{Y} \subseteq \mathcal{T}$ with $\boldsymbol{Y}$ finite and $\mathcal{T}$ is a topology on S, $\bigcap \boldsymbol{Y} \in \mathcal{T}$. Since $\boldsymbol{Y} \subseteq \mathcal{U}$ with $\boldsymbol{Y}$ finite and $\mathcal{U}$ is a topology on S, $\bigcap \boldsymbol{Y} \in \mathcal{U}$. Since $\bigcap \boldsymbol{Y} \in \mathcal{T}$ and $\bigcap \boldsymbol{Y} \in \mathcal{U}$, we have $\bigcap \boldsymbol{Y} \in \mathcal{T} \cap \mathcal{U}$.

Therefore, $\mathcal{T} \cap \mathcal{U}$ is a topology on S. □

2. Let $(S, \mathcal{T})$ be a topological space and let A be a subspace of S.
   (i) Provide an example to show that if B is open in A in the subspace topology, then B need not be open in S.
   (ii) Provide an example to show that if B is closed in A in the subspace topology, then B need not be closed in S.

Solutions:

(i) Let $S = \mathbb{R}$ and let $\mathcal{T}$ be the standard topology on $\mathbb{R}$. Let $A = [0, 1]$ and let $B = \left[0, \frac{1}{2}\right)$. Then we have $B = \left(-1, \frac{1}{2}\right) \cap [0, 1]$. Therefore, B is open in the subspace topology of A. However, B is **not** open in the standard topology of $\mathbb{R}$.

(ii) Let $S = \mathbb{R}$ and let $\mathcal{T}$ be the standard topology on $\mathbb{R}$. Let $A = [0, 1)$ and let $B = \left[\frac{1}{2}, 1\right)$. Then we have $A \setminus B = \left[0, \frac{1}{2}\right) = \left(-1, \frac{1}{2}\right) \cap [0, 1)$. So, $A \setminus B$ is open in the subspace topology of A. Therefore, B is closed in the subspace topology of A. However, B is **not** closed in the standard topology of $\mathbb{R}$.

LEVEL 2

3. Let $K = \left\{\frac{1}{n} \mid n \in \mathbb{Z}^+\right\}$, $\mathcal{B} = \{(a, b) \mid a, b \in \mathbb{R} \land a < b\} \cup \{(a, b) \setminus K \mid a, b \in \mathbb{R} \land a < b\}$. Prove that $\mathcal{B}$ is a basis for a topology $\mathcal{T}_K$ on $\mathbb{R}$.

Proof: We already know that the bounded open intervals alone cover $\mathbb{R}$. So, $\mathcal{B}$ covers $\mathbb{R}$. By Problem 5 from Problem Set 7 (part (ii)), the intersection of two open intervals with nonempty intersection is an open interval. Furthermore, we have $(a, b) \cap [(c, d) \setminus K] = [(a, b) \cap (c, d)] \setminus K$ and we have $[(a, b) \setminus K] \cap [(c, d) \setminus K] = [(a, b) \cap (c, d)] \setminus K$. This shows that $\mathcal{B}$ has the intersection containment property. Therefore, $\mathcal{B}$ is a basis for a topology $\mathcal{T}_K$ on $\mathbb{R}$. □

4. Prove that $\mathcal{B} = \{X \subseteq \mathbb{R} \mid \mathbb{R} \setminus X \text{ is finite}\}$ generates a topology $\mathcal{T}$ on $\mathbb{R}$ that is strictly coarser than the standard topology. $\mathcal{T}$ is called the **cofinite topology** on $\mathbb{R}$.

Proof: Let $x \in \mathbb{R}$. Then $X = \mathbb{R} \setminus \{x + 1\} \in \mathcal{B}$ because $\mathbb{R} \setminus X = \mathbb{R} \setminus (\mathbb{R} \setminus \{x + 1\}) = \{x + 1\}$, which is finite and $x \in \mathbb{R} \setminus \{x + 1\}$ because $x \neq x + 1$. So, $\mathcal{B}$ covers $\mathbb{R}$. Let $x \in \mathbb{R}$, and let $X, Y \in \mathcal{B}$ with $x \in X \cap Y$. Then $\mathbb{R} \setminus X$ and $\mathbb{R} \setminus Y$ are both finite and $\mathbb{R} \setminus (X \cap Y) = (\mathbb{R} \setminus X) \cup (\mathbb{R} \setminus Y)$ (by De Morgan's Law) is the union of two finite sets, thus finite. It follows that $X \cap Y \in \mathcal{B}$. Therefore, $\mathcal{B}$ has the intersection containment property. Since $\mathcal{B}$ covers $\mathbb{R}$ and has the intersection containment property, $\mathcal{B}$ is a basis for a topology $\mathcal{T}$ on $\mathbb{R}$.

Note that the topology generated by $\mathcal{B}$ is simply $\mathcal{B} \cup \{\emptyset\}$. $\mathbb{R}$ is in the basis $\mathcal{B}$ because $\mathbb{R} \setminus \mathbb{R} = \emptyset$, which of course is finite. If $\boldsymbol{X} \subseteq \mathcal{B}$, then $\mathbb{R} \setminus \bigcup \boldsymbol{X} = \bigcap\{\mathbb{R} \setminus A \mid A \in \boldsymbol{X}\}$. This is an intersection of finite sets, which is finite. So, $\bigcup \boldsymbol{X} \in \mathcal{B}$. Finally, if $\boldsymbol{Y} \subseteq \mathcal{T}$ and $\boldsymbol{Y}$ is finite, then $\mathbb{R} \setminus \bigcap \boldsymbol{Y} = \bigcup\{\mathbb{R} \setminus A \mid A \in \boldsymbol{Y}\}$. This is a finite union of finite sets, which is finite. So, $\bigcap \boldsymbol{Y} \in \mathcal{B}$.

It follows that $\mathcal{T}$, the topology generated by $\mathcal{B}$, consists of only the cofinite sets together with the empty set.

Since $(0, 1)$ is open in the standard topology of $\mathbb{R}$ and is **not** cofinite, $(0, 1)$ is **not** open in the cofinite topology. Therefore, the cofinite topology on $\mathbb{R}$ is strictly coarser than the standard topology on $\mathbb{R}$. □

5. Let S be a nonempty set and let $\mathcal{T} = \{X \subseteq S \mid S \setminus X \text{ is countable}\} \cup \{\emptyset\}$. Is $\mathcal{T}$ a topology on S?

Solution: We show that $\mathcal{T}$ is a topology on S. $\emptyset \in \mathcal{T}$ by definition. $S \setminus S = \emptyset$, which is countable, and so, $S \in \mathcal{T}$.

Let $\boldsymbol{X} \subseteq \mathcal{T}$. Then $\boldsymbol{X} \subseteq \mathcal{T}$. Then $S \setminus A$ is countable for each $A \in \boldsymbol{X}$. Now, $S \setminus \bigcup \boldsymbol{X} = \bigcap\{S \setminus A \mid A \in \boldsymbol{X}\}$. Since this is countable, we have $\bigcup \boldsymbol{X} \in \mathcal{T}$.

Let $\boldsymbol{Y} \subseteq \mathcal{T}$ with $\boldsymbol{Y}$ finite. Then $S \setminus A$ is countable for each $A \in \boldsymbol{Y}$. Now, $S \setminus \bigcap \boldsymbol{Y} = \bigcup\{S \setminus A \mid A \in \boldsymbol{Y}\}$. This is a finite union of countable sets, which is countable. So, $\bigcap \boldsymbol{X} \in \mathcal{T}$.

It follows that $\mathcal{T}$ is a topology on S. □

6. Let $\mathcal{X}$ be a set of topologies on a set S. Prove that $\bigcap \mathcal{X}$ is a topology on S.

Proof: Let $\mathcal{T} \in \mathcal{X}$. Since $\mathcal{T}$ is a topology on S, $\emptyset \in \mathcal{T}$ and $S \in \mathcal{T}$. Since $\mathcal{T} \in \mathcal{X}$ was arbitrary, we have $\emptyset \in \bigcap\mathcal{X}$ and $S \in \bigcap\mathcal{X}$.

Let $\boldsymbol{X} \subseteq \bigcap\mathcal{X}$. Then $\boldsymbol{X} \subseteq \mathcal{T}$ for each $\mathcal{T} \in \mathcal{X}$. So, $\bigcup\boldsymbol{X} \in \mathcal{T}$ for each $\mathcal{T} \in \mathcal{X}$. Therefore, $\bigcup\boldsymbol{X} \in \bigcap\mathcal{X}$.

Let $\boldsymbol{Y} \subseteq \bigcap\mathcal{X}$ with $\boldsymbol{Y}$ finite. Then $\boldsymbol{Y} \subseteq \mathcal{T}$ for each $\mathcal{T} \in \mathcal{X}$. So, $\bigcap\boldsymbol{Y} \in \mathcal{T}$ for each $\mathcal{T} \in \mathcal{X}$. It follows that $\bigcap\boldsymbol{Y} \in \bigcap\mathcal{X}$.

Therefore, $\bigcap\mathcal{X}$ is a topology on S. □

LEVEL 3

7. Let $(S, \mathcal{T})$ be a topological space and let A be a subspace of S.
   (i) Prove that if B is open in A in the subspace topology and A is open in S, then B is open in S.
   (ii) Prove that B is closed in A if and only if $B = C \cap A$ for some closed set C of S.
   (iii) Prove that if B is closed in A in the subspace topology and A is closed in S, then B is closed in S.

Proofs:

(i) Suppose that B is open in A in the subspace topology and A is open in S. Since B is open in A, there is $U \in \mathcal{T}$ such that $B = U \cap A$. Since A is open in S, we have $A \in \mathcal{T}$. Since $U \in \mathcal{T}$, $A \in \mathcal{T}$, and $\mathcal{T}$ is a topology on S, $B = U \cap A \in \mathcal{T}$. So, B is open in S. □

(ii) First suppose that B is closed in A. Then $A \setminus B$ is open in A. Therefore, there is $U \in \mathcal{T}$ such that $A \setminus B = U \cap A$. Let $C = S \setminus U$. Then C is closed in S. We now show that $B = C \cap A$. To see this, first let $x \in B$. Then $x \notin A \setminus B$. So, $x \notin U \cap A$. Since $x \in A$ (because $B \subseteq A$), $x \notin U$. Therefore, $x \in S \setminus U = C$. Since $x \in C$ and $x \in A$, we have $x \in C \cap A$. Next, let $x \in C \cap A$. Then $x \in C$ and $x \in A$. Since $x \in C$ and $C = S \setminus U$, $x \notin U$. So, we have $x \in A$ and $x \notin U$. It follows that $x \in A \setminus U$. Therefore, $x \notin U \cap A = A \setminus B$. Since $x \in A$ and $x \notin A \setminus B$, we have $x \in B$. Therefore, $B = C \cap A$, as desired.

Conversely, suppose that $B = C \cap A$ for some closed set C of S. Let $U = S \setminus C$. Then U is open in S. We now show that $A \setminus B = U \cap A$. To see this, first let $x \in A \setminus B$. Then $x \in A$ and $x \notin B$. Since, $x \in A$ and $x \notin B = C \cap A$, we have $x \notin C$. So, $x \in S \setminus C = U$. Since $x \in U$ and $x \in A$, we have $x \in U \cap A$. Next, let $x \in U \cap A$. Then $x \in U = S \setminus C$ and $x \in A$. So, $x \notin C$. Since $x \notin C$ and $C \cap A \subseteq C$, we have $x \notin C \cap A = B$. Since $x \in A$ and $x \notin B$, we have $x \in A \setminus B$. Since $A \setminus B = U \cap A$ with U open in S, $A \setminus B$ is open in A. Therefore, B is closed in A. □

(iii) Suppose that B is closed in A in the subspace topology and A is closed in S. Since B is closed in A, by (ii), there is a closed set C of S such that $B = C \cap A$. Since C and A are closed in S, $S \setminus C$ and $S \setminus A$ are open in S. Since $S \setminus C \in \mathcal{T}$, $S \setminus A \in \mathcal{T}$, and $\mathcal{T}$ is a topology on S, we have that $S \setminus B = S \setminus (C \cap A) = (S \setminus C) \cup (S \setminus A) \in \mathcal{T}$ (by De Morgan's Law). So, $S \setminus B$ is open in S, and therefore, B is closed in S. □

8. Let $(S, \mathcal{T})$ be a topological space and let X and Y be subsets of S. Prove that $\overline{X \cup Y} = \overline{X} \cup \overline{Y}$.

Proof: Let X and Y be subsets of S and let $x \in \overline{X \cup Y}$. By part 3 of Theorem 9.4, $x \in X \cup Y$ or x is an accumulation point of $X \cup Y$. If $x \in X \cup Y$, then $x \in X$ or $x \in Y$. By part 1 of Theorem 9.4, $X \subseteq \overline{X}$ and $Y \subseteq \overline{Y}$. So, $x \in \overline{X}$ or $x \in \overline{Y}$. Thus, $x \in \overline{X} \cup \overline{Y}$. Now, suppose that x is an accumulation point of $X \cup Y$, but x is **not** an accumulation point of Y. Then there is an open set U with $x \in U$ and such that if $y \in Y$ with $y \neq x$, then $y \notin U$. Let V be any open set containing x and let $W = U \cap V$. Since $x \in W$, W is a nonempty open set. Since x is an accumulation point of $X \cup Y$, there is $y \in X \cup Y$ with $y \in W$ and $y \neq x$. Since $W \subseteq U$, $y \notin Y$. Therefore, $y \in X$. Since $W \subseteq V$, we have $y \in V$. It follows that x is an accumulation point of X. So, $x \in \overline{X}$. Therefore, $x \in \overline{X}$ or $x \in \overline{Y}$, and so, $x \in \overline{X} \cup \overline{Y}$. Since $x \in \overline{X \cup Y}$ was arbitrary, we have $\overline{X \cup Y} \subseteq \overline{X} \cup \overline{Y}$.

Conversely, let $x \in \overline{X} \cup \overline{Y}$. Then $x \in \overline{X}$ or $x \in \overline{Y}$. Without loss of generality, assume that $x \in \overline{X}$. By part 3 of Theorem 9.4, $x \in X$ or x is an accumulation point of X. If $x \in X$, then $x \in X$ or $x \in Y$. So, $x \in X \cup Y$. By part 1 of Theorem 9.4, $X \cup Y \subseteq \overline{X \cup Y}$. So, $x \in \overline{X \cup Y}$. Now, suppose that x is an accumulation point of X and let U be any open set containing x. Then there is $y \in X$ with $y \in U$ and $y \neq x$. So, $y \in X$ or $y \in Y$, and thus, $y \in X \cup Y$. It follows that x is an accumulation point of $X \cup Y$. Therefore, $x \in \overline{X \cup Y}$. Since $x \in \overline{X} \cup \overline{Y}$ was arbitrary, $\overline{X} \cup \overline{Y} \subseteq \overline{Z}$.

Since $\overline{Z} \subseteq \overline{X} \cup \overline{Y}$ and $\overline{X} \cup \overline{Y} \subseteq \overline{Z}$, we have $\overline{Z} = \overline{X} \cup \overline{Y}$. □

9. Let $(S, \mathcal{T})$ be a topological space and let $A \subseteq B$ with A and B be subspaces of S. Prove that the subspace topology of A with respect to S is the same as the subspace topology of A with respect to B.

Proof: Let U be open in the subspace topology of A with respect to S. Then $U = V \cap A$ for some set V that is open in S. Since V is open in S, $W = V \cap B$ is open in B. We have

$$W \cap A = (V \cap B) \cap A = V \cap (B \cap A) = V \cap A = U.$$

Therefore, U is open in the subspace topology of A with respect to B.

Conversely, let U be open in the subspace topology of A with respect to B. Then $U = V \cap A$ for some set V that is open in B. Then $V = W \cap B$ for some set W that is open in S. We have

$$U = V \cap A = (W \cap B) \cap A = W \cap (B \cap A) = W \cap A.$$

Therefore, U is open in the subspace topology of A with respect to S. □

10. Let S be a nonempty set and let $\mathcal{B}$ be a collection of subsets of S. Prove that the set generated by $\mathcal{B}$, $\{\bigcup \boldsymbol{X} \mid \boldsymbol{X} \subseteq \mathcal{B}\}$, is equal to $\{A \subseteq S \mid \forall x \in A\ \exists B \in \mathcal{B}(x \in B \wedge B \subseteq A)\}$.

Proof: Let $\mathcal{C} = \{\bigcup \boldsymbol{X} \mid \boldsymbol{X} \subseteq \mathcal{B}\}$ and let $\mathcal{D} = \{A \subseteq S \mid \forall x \in A\ \exists B \in \mathcal{B}(x \in B \wedge B \subseteq A)\}$. First, let $A \in \mathcal{C}$. Then there is $\boldsymbol{X} \subseteq \mathcal{B}$ such that $A = \bigcup \boldsymbol{X}$. Let $x \in A$. Then there is a $B \in \boldsymbol{X}$ with $x \in B$. Since $\boldsymbol{X} \subseteq \mathcal{B}$, $B \in \mathcal{B}$. Also, since $B \in \boldsymbol{X}$, $B \subseteq \bigcup \boldsymbol{X} = A$. Therefore, $A \in \mathcal{D}$. Since $A \in \mathcal{C}$ was arbitrary, $\mathcal{C} \subseteq \mathcal{D}$.

Now, let $A \in \mathcal{D}$. For each $x \in A$, there is $B_x \in \mathcal{B}$ such that $x \in B_x$ and $B_x \subseteq A$. If $y \in A$, then $y \in B_y$. So, $y \in \bigcup\{B_x \mid x \in A\}$. So, $A \subseteq \bigcup\{B_x \mid x \in A\}$. If $y \in \bigcup\{B_x \mid x \in A\}$, then $y \in B_x$ for some $x \in A$. Since $B_x \subseteq A$, $y \in A$. Therefore, $\bigcup\{B_x \mid x \in A\} \subseteq A$. It follows that $A = \bigcup\{B_x \mid x \in A\}$. In other words, we have $A = \bigcup \boldsymbol{X}$, where $\boldsymbol{X} = \{B_x \mid x \in A\}$. Since $A \in \mathcal{D}$ was arbitrary, $\mathcal{D} \subseteq \mathcal{C}$.

Since $\mathcal{C} \subseteq \mathcal{D}$ and $\mathcal{D} \subseteq \mathcal{C}$, we have $\mathcal{C} = \mathcal{D}$. □

LEVEL 4

11. Let $(S, \mathcal{T})$ be a topological space. Prove that an arbitrary intersection of closed sets in S is a closed set in S and a finite union of closed sets in S is a closed set in S.

Proof: Let $\boldsymbol{X}$ be a nonempty set of closed sets in S. Then for each $A \in \boldsymbol{X}$, $S \setminus A$ is an open set in S. Since $(S, \mathcal{T})$ is a topological space, $\bigcup\{S \setminus A | A \in \boldsymbol{X}\}$ is open in S. Therefore, $S \setminus \bigcup\{S \setminus A | A \in \boldsymbol{X}\}$ is closed in S. By De Morgan' s Law, $\bigcap \boldsymbol{X} = S \setminus \bigcup\{S \setminus A | A \in \boldsymbol{X}\}$. Therefore, $\bigcap \boldsymbol{X}$ is closed in S.

Now, let $\boldsymbol{X}$ be a finite set of closed subsets of S. Then for each $A \in \boldsymbol{X}$, $S \setminus A$ is an open set in S. Since $(S, \mathcal{T})$ is a topological space, $\bigcap\{S \setminus A | A \in \boldsymbol{X}\}$ is open in S. Therefore, $S \setminus \bigcap\{S \setminus A | A \in \boldsymbol{X}\}$ is closed in S. By De Morgan' s Law, $\bigcup \boldsymbol{X} = S \setminus \bigcap\{S \setminus A | A \in \boldsymbol{X}\}$. Therefore, $\bigcup \boldsymbol{X}$ is closed in S. □

12. Let $\mathcal{B}' = \{(a, b) \mid a, b \in \mathbb{Q} \wedge a < b\}$. Prove that $\mathcal{B}'$ is countable and that $\mathcal{B}'$ is a basis for a topology on $\mathbb{R}$. Then show that the topology generated by $\mathcal{B}'$ is the standard topology on $\mathbb{R}$.

Proof: Define $g: \mathcal{B}' \to \mathbb{Q} \times \mathbb{Q}$ by $g\big((a, b)\big) = (a, b)$ (the open interval (a, b) is being sent to the ordered pair (a, b)—it is unfortunate that the notation for these two objects is identical). If $g\big((a, b)\big) = g\big((c, d)\big)$, then $(a, b) = (c, d)$ (as ordered pairs). So, $a = c$ and $b = d$. Therefore, $(a, b) = (c, d)$ (as open intervals). This shows that g is injective. So, $\mathcal{B}' \preccurlyeq \mathbb{Q} \times \mathbb{Q} \sim \mathbb{Q}$. Since $\mathbb{Q}$ is countable, so is $\mathcal{B}'$.

If $x \in \mathbb{R}$, then $x - 1 < x < x + 1$. By the Density Theorem, we can choose $a, b \in \mathbb{Q}$ such that $x - 1 < a < x$ and $x < b < x + 1$. Then $x \in (a, b)$ and $(a, b) \in \mathcal{B}'$. So, $\mathcal{B}'$ covers $\mathbb{R}$. Now, let $x \in \mathbb{R}$ and $(a, b), (c, d) \in \mathcal{B}'$ with $x \in (a, b) \cap (c, d)$. By Problem 5 from Problem Set 7 (part (ii)), we have $(a, b) \cap (c, d) = (e, f)$ for some $e, f \in \mathbb{R}$. By the Density Theorem, we can choose $g, h \in \mathbb{Q}$ such that $e < g < x$ and $x < h < f$. Then $x \in (g, h)$ and $(g, h) \subseteq (a, b) \cap (c, d)$. So, $\mathcal{B}'$ has the intersection containment property. It follows that $\mathcal{B}'$ is a basis for a topology on $\mathbb{R}$.

Since every open interval with rational endpoints is open in the standard topology on $\mathbb{R}$, the topology generated by $\mathcal{B}'$ is contained in the standard topology. Let $a, b \in \mathbb{R}$. For each $n \in \mathbb{Z}^+$, by the Density Theorem, we can choose $q_n, r_n \in \mathbb{Q}$ with $a < q_n < a + \frac{1}{n}$ and $b - \frac{1}{n} < r_n < b$. We will now show that $(a, b) = \bigcup\{(q_n, r_n) \mid n \in \mathbb{Z}^+\}$. If $x \in (a, b)$, then there is $n \in \mathbb{Z}^+$ with $x \in \left(a + \frac{1}{n}, b - \frac{1}{n}\right) \subseteq (q_n, r_n)$. So, $x \in \bigcup\{(q_n, r_n) \mid n \in \mathbb{Z}^+\}$. Therefore, $(a, b) \subseteq \bigcup\{(q_n, r_n) \mid n \in \mathbb{Z}^+\}$. If $x \in \bigcup\{(q_n, r_n) \mid n \in \mathbb{Z}^+\}$, then there is $n \in \mathbb{Z}^+$ such that $x \in (q_n, r_n) \subseteq (a, b)$. So, $\bigcup\{(q_n, r_n) \mid n \in \mathbb{Z}^+\} \subseteq (a, b)$. It follows that $(a, b) = \bigcup\{(q_n, r_n) \mid n \in \mathbb{Z}^+\}$. Since (a, b) is a union of sets in $\mathcal{B}'$, the standard topology is contained in the topology generated by $\mathcal{B}'$. So, $\mathcal{B}'$ generates the standard topology on $\mathbb{R}$. □

13. Let $(S, \mathcal{T})$ be a topological space, let A be a subspace of S, let B be a subset of A, and let $\overline{B}$ be the closure of B in S. Prove that the closure of B in A is $\overline{B} \cap A$.

Proof: Let Y be the closure of B in A. We want to show that $Y = \overline{B} \cap A$. Since $\overline{B}$ is closed in S, by part (ii) of Problem 7, $\overline{B} \cap A$ is closed in A. Since $B \subseteq \overline{B}$ and $B \subseteq A$, we have $B \subseteq \overline{B} \cap A$. By definition, Y is the intersection of all closed subsets of A containing B, we have $Y \subseteq \overline{B} \cap A$.

Also, Y is closed in A. So, by part (ii) of Problem 7, $Y = C \cap A$ for some set C that is closed in S. Then C is a closed set of S containing B. Since $\overline{B}$ is the intersection of all closed sets in S thst contain B, we see that $\overline{B} \subseteq C$. So, $\overline{B} \cap A \subseteq C \cap A = Y$.

Since $Y \subseteq \overline{B} \cap A$ and $\overline{B} \cap A \subseteq Y$, we have $Y = \overline{B} \cap A$. □

14. Let T_L be the set generated by the half open intervals of the form $[a, b)$ with $a, b \in \mathbb{R}$. Show that T_L is a topology on $\mathbb{R}$ that is strictly finer than the standard topology on $\mathbb{R}$ and incomparable with the topology $\mathcal{T}_K$ from Problem 3 above.

Proof: Let $\mathcal{B} = \{[a, b) \mid a, b \in \mathbb{R}\}$ and let $\mathcal{T}_L$ be the set generated by $\mathcal{B}$. Let $x \in \mathbb{R}$. Then $x \in [x, x+1)$. This shows that $\mathcal{B}$ covers $\mathbb{R}$. If $[a, b), [c, d) \in \mathcal{B}$ with $[a, b) \cap [c, d) \neq \emptyset$, then $[a, b) \cap [c, d) = [e, f)$, where $e = \max\{a, c\}$ and $f = \min\{b, d\}$. To see this, let $x \in [a, b) \cap [c, d)$. Then we have $a \leq x < b$ and $c \leq x < d$. Since $a \leq x$ and $c \leq x$, $e \leq x$. Since $x < b$ and $x < d$, $x < f$. It follows that $x \in [e, f)$. Conversely, if $x \in [e, f)$, then $e \leq x < f$. Since $a \leq e$ and $c \leq e$, $a \leq x$ and $c \leq x$. Since $f \leq b$ and $f \leq d$, $x < b$ and $x < d$. So, $x \in [a, b)$ and $x \in [c, d)$. Thus, $x \in [a, b) \cap [c, d)$. It follows that $\mathcal{B}$ has the intersection containment property. Since $\mathcal{B}$ covers $\mathbb{R}$ and $\mathcal{B}$ has the intersection containment property, $\mathcal{B}$ is a basis for a topology on $\mathbb{R}$.

To see that $\mathcal{T}_L$ is finer than the standard topology on $\mathbb{R}$, note that each basic open set (a, b) in the standard topology is equal to the union $\bigcup\left\{\left[a + \frac{1}{n}, b\right) \middle| n \in \mathbb{Z}^+\right\}$. See Problem 8 from Problem Set 6 for a proof similar to what is needed to prove this result.

To see that $\mathcal{T}_L$ is **strictly** finer than the standard topology, just note that $[0, 1)$ cannot be written as a union of bounded open intervals, for 0 would need to be inside one of those open intervals, and it would then follow that there is an $x < 0$ with $x \in [0,1)$.

The set $(-1,1) \setminus K$ is open in $\mathcal{T}_K$. We show that $(-1,1) \setminus K$ is **not** open in $\mathcal{T}_L$. If $(-1,1) \setminus K$ is the union of sets of the form $[a, b)$, then 0 would need to be inside one of those half-open intervals, let's say that $0 \in [a, b)$. But then there is some $n > 0$ such that $\frac{1}{n} < b$ (use the Archimedean property). Therefore, $\frac{1}{n} \in [a, b)$. This contradicts that $\frac{1}{n} \in K$. This shows that $\mathcal{T}_L$ is **not** finer than $\mathcal{T}_K$.

The set $[0,1)$ is open in $\mathcal{T}_L$. We've already seen that $[0,1)$ cannot be written as a union of bounded open intervals. If we throw additional sets of the form $(a, b) \setminus K$ into such a union, then we still run into the same issue with 0. If $0 \in (a, b)$ or $(a, b) \setminus K$, we would get an $x < 0$ with $x \in [0,1)$. □

LEVEL 5

15. Let K be an index set and let $(S_k, \mathcal{T}_k)$ be a topological space for each $k \in K$. Prove that the collection $\mathcal{B} = \{\prod U_k \mid \forall k \in K(U_k \in \mathcal{T}_k) \wedge (U_k = S_k \text{ for all but finitely many } k)\}$ is a basis for a topology $\mathcal{T}$ on $\prod S_k$. Furthermore, if $\mathcal{B}_k$ is a basis for $\mathcal{T}_k$ for each $k \in K$, then the collection $\mathcal{C} = \{\prod U_k \mid \forall k \in K(U_k \in \mathcal{B}_k) \wedge (U_k = S_k \text{ for all but finitely many } k)\}$ is a basis for the same topology (this topology is called the **product topology** on $\prod S_k$). Show that in both cases, if we remove "$U_k = S_k$ for all but finitely many k," we get a topology (called the **box topology**) that is finer than the product topology.

Proof: Let $(s_k) \in \prod S_k$. Since $S_k \in \mathcal{T}_k$ for each $k \in K$, $\prod S_k \in \mathcal{B}$. So, $\mathcal{B}$ covers $\prod S_k$.

Now, let $(s_k) \in \prod U_k \cap \prod V_k$, where $\prod U_k, \prod V_k \in \mathcal{B}$. Since $U_k, V_k \in \mathcal{T}_k$ for each $k \in K$, $U_k \cap V_k \in \mathcal{T}_k$ for each $k \in K$. Also, since $U_k = S_k$ for all but finitely many k and $V_k = S_k$ for all but finitely many k, it follows that $U_k \cap V_k = S_k$ for all but finitely many k. Therefore, $\prod(U_k \cap V_k) \in \mathcal{B}$. Since $\prod U_k \cap \prod V_k = \prod(U_k \cap V_k)$ (Check this!), we have $\prod U_k \cap \prod V_k \in \mathcal{B}$. So, $\mathcal{B}$ has the intersection containment property.

Since $\mathcal{B}$ covers $\prod S_k$ and $\mathcal{B}$ has the intersection containment property, it follows that $\mathcal{B}$ is a basis for a topology on $\prod S_k$.

Since $\mathcal{B}_k \subseteq \mathcal{T}_k$ for each $k \in K$, $\mathcal{C} \subseteq \mathcal{B}$. Therefore, the set generated by $\mathcal{C}$ is contained in the set generated by $\mathcal{B}$, which is $\mathcal{T}$.

Now, let $\prod U_k \in \mathcal{B}$ and let $(x_k) \in \prod U_k$ Then $x_k \in U_k$ for each $k \in K$. Since for each $k \in K$, $U_k \in \mathcal{T}_k$, for each $k \in K$, there is $U_k^{x_k} = \mathcal{B}_k$ with $U_k^{x_k} \subseteq U_k$. Let $W = \bigcup\{\prod U_k^{x_k} \mid x_k \in U_k \text{ for each } k \in K\}$. Then $\prod U_k = W$ (Check this!). Therefore, $\prod U_k$ is a union of elements from $\mathcal{C}$. Thus, $\mathcal{T}$ (the set generated by $\mathcal{B}$) is contained in the set generated by $\mathcal{C}$. It follows that $\mathcal{B}$ and $\mathcal{C}$ generate the same set. Since $\mathcal{B}$ generates $\mathcal{T}$, so does $\mathcal{C}$. Therefore, $\mathcal{C}$ is a basis for $\mathcal{T}$.

Let $\mathcal{D} = \{\prod U_k \mid \forall k \in K(U_k \in \mathcal{T}_k)\}$. An argument nearly identical to the one given in the first three paragraphs above shows that $\mathcal{D}$ is a basis for a topology on $\prod S_k$. Clearly $\mathcal{D} \subseteq \mathcal{B}$, and therefore, the topology generated by D is finer than $\mathcal{T}$. □

16. Let K be an index set, let $(S_k, \mathcal{T}_k)$ be a topological space for each $k \in K$, and let $A_k \subseteq S_k$ for each $k \in K$. Prove that $\overline{\prod A_k} = \prod \overline{A_k}$.

Proof: First note that for any set X in a topological space, $x \in \overline{X}$ if and only if for every open set U containing x, we have $U \cap X \neq \emptyset$.

To see this, first assume that $x \in \overline{X}$ and let U be an open set containing x. If $x \in X$, then $x \in U \cap X$. So, assume that $x \notin X$. Then by part 3 of Theorem 9.4, x is an accumulation point of X. So, there is $y \in U \cap X$ with $y \neq x$.

Conversely, assume that for every open set U containing x, $U \cap X \neq \emptyset$. If $x \in X$, then since $X \subseteq \overline{X}$, we have $x \in \overline{X}$. So, assume that $x \notin X$. Let U be an open set containing x. Since $U \cap X \neq \emptyset$, there is $y \in U \cap X$ with $y \neq x$. It follows that x is an accumulation point of X. By part 3 of Theorem 9.4, $x \in \overline{X}$.

Now, let $(x_k) \in \overline{\prod A_k}$. We will show that for each $n \in K$, $x_n \in \overline{A_n}$. To see this, let V_n be an open set contining x_n. Let $U = \prod U_k$, where $U_n = V_n$ and $U_k = S_k$ for all $k \neq n$. Then U is an open set in $\prod S_k$ containing (x_k). Since $(x_k) \in \overline{\prod A_k}$, there is $(y_k) \in U \cap \prod A_k$. Then $y_n \in V_n \cap A_n$. Therefore, $x_n \in \overline{A_n}$. Since $n \in K$ was arbitrary, we see that $(x_k) \in \prod \overline{A_k}$. It follows that $\overline{\prod A_k} \subseteq \prod \overline{A_k}$.

Now, let $(x_k) \in \prod \overline{A_k}$ and for each $k \in K$, let $U_k \in \mathcal{T}_k$ with $U_k = S_k$ for all but finitely many k. Then $U = \prod U_k$ is a basis element for the product topology on $\prod S_k$. Let $k \in K$. Since $x_k \in \overline{A_k}$, we can choose $y_k \in U_k \cap A_k$. Then $(y_k) \in U \cap \prod A_k$. Thus, $(x_k) \in \overline{\prod A_k}$. It follows that $\prod \overline{A_k} \subseteq \overline{\prod A_k}$.

Since $\overline{\prod A_k} \subseteq \prod \overline{A_k}$ and $\prod \overline{A_k} \subseteq \overline{\prod A_k}$, we have $\overline{\prod A_k} \subseteq \prod \overline{A_k}$. □

17. Let J and K be index sets. Prove that $\bigcap\{\prod_{k \in K} B_k^j \mid j \in J\} = \prod_{k \in K}(\bigcap\{B_k^j \mid j \in J\})$.

Proof: $\mathbf{x} \in \bigcap\{\prod_{k \in K} B_k^j \mid j \in J\}$ if and only if $\forall j \in J(\mathbf{x} \in \prod_{k \in K} B_k^j)$ if and only $\forall j \in J\ \forall k \in K(x_k \in B_k^j)$ if and only if $\forall k \in K\ \forall j \in J(x_k \in B_k^j)$ if and only if $\forall k \in K\ (x_k \in \bigcap\{B_k^j \mid j \in J\})$ if and only if $\mathbf{x} \in \prod_{k \in K}(\bigcap\{B_k^j \mid j \in J\})$ □

Problem Set 10

LEVEL 1

1. Let $(S, \mathcal{T})$ be a T_1-space with $|S| \geq 2$ and let $\mathcal{B}$ be a basis for $(S, \mathcal{T})$. Prove that $\mathcal{B} \setminus \{S\}$ is a basis for $(S, \mathcal{T})$.

Proof: Let $(S, \mathcal{T})$ be a T_1-space with $|S| \geq 2$ and let $\mathcal{B}$ be a basis for $(S, \mathcal{T})$. We need to show that $\mathcal{B} \setminus \{S\}$ covers S. Let $x \in S$. Since $|S| \geq 2$, there is $y \in S$ with $x \neq y$. Since $(S, \mathcal{T})$ is a T_1-space, there is an open set U with $x \in U$ and $y \notin U$. Since U is a union of basis elements, there is $V \in \mathcal{B}$ with $x \in V$ and $V \subseteq U$. Since $y \notin U$, it follows that $y \notin V$. So, $V \neq S$. Since $x \in S$ was arbitrary, $\mathcal{B} \setminus \{S\}$ covers S. Therefore, $\mathcal{B} \setminus \{S\}$ is a basis for $(S, \mathcal{T})$. □

2. Prove that a closed subspace of a T_4-space is a T_4-space.

Proof: Let $(S, \mathcal{T})$ be a T_4-space, let $(C, \mathcal{T}_C)$ be a closed subspace of $(S, \mathcal{T})$, and let A, B be disjoint closed subsets of C. Since A is closed in C and C is closed in S, by part (iii) of Problem 7 from Problem Set 9, A is closed in S. Similarly, B is closed in S. Since $(S, \mathcal{T})$ be a T_4-space, there are open sets U and V with $A \subseteq U$, $B \subseteq V$, and $U \cap V = \emptyset$. $U \cap C$ and $V \cap C$ are open in the subspace topology, $A \subseteq U \cap C$, $B \subseteq V \cap C$, and $(U \cap C) \cap (V \cap C) = (U \cap V) \cap C = \emptyset \cap C = \emptyset$. □

LEVEL 2

3. Let $(S, \mathcal{T})$ be a T_1-space. Prove that $(S, \mathcal{T})$ is a T_3-space if and only if for every point $x \in S$ and every open set U containing x, there is an open set V containing x such that $\overline{V} \subseteq U$.

Proof: Suppose that $(S, \mathcal{T})$ is a T_3-space, let $x \in S$ and let U be an open set containing x. Then the set $A = S \setminus U$ is closed in S. Since $(S, \mathcal{T})$ is a T_3-space, there are disjoint open sets V and W such that $x \in V$ and $A \subseteq W$. If $y \in A$, then W is an open set disjoint from V with $y \in W$. So, $y \notin \overline{V}$. Therefore, $y \in \overline{V}$ implies that $y \notin A$, or equivalently, $y \in U$. Since $y \in \overline{V}$ was arbitrary, $\overline{V} \subseteq U$.

Conversely, let $(S, \mathcal{T})$ is a T_1-space such that for every point $x \in S$ and every open set U containing x, there is an open set V containing x such that $\overline{V} \subseteq U$. Let $x \in S$ and let A be a closed set in S not containing x. Then $U = S \setminus A$ is an open set containing x. So, by our assumption, there is an open set W containing x such that $\overline{W} \subseteq U$. Then W and $S \setminus \overline{W}$ are open sets, $x \in W$, $A \subseteq S \setminus \overline{W}$ (if $x \in A$, then $x \notin S \setminus A = U$, and so, $x \notin \overline{W}$). Also, $W \cap (S \setminus \overline{W}) \subseteq W \cap (S \setminus W) = \emptyset$. □

4. Let $(S, \mathcal{T})$ be a T_1-space. Prove that $(S, \mathcal{T})$ is a T_4-space if and only if for any closed set A in S and every open set U containing A, there is an open set V containing A such that $\overline{V} \subseteq U$.

Proof: Suppose that $(S, \mathcal{T})$ is a T_4-space, let A be closed in S, and let U be an open set containing A. Then the set $B = S \setminus U$ is closed in S. Since $(S, \mathcal{T})$ is a T_4-space, there are disjoint open sets V and W such that $A \subseteq V$ and $B \subseteq W$. If $y \in B$, then W is an open set disjoint from V with $y \in W$. So, $y \notin \overline{V}$. Therefore, $y \in \overline{V}$ implies that $y \notin B$, or equivalently, $y \in U$. Since $y \in \overline{V}$ was arbitrary, $\overline{V} \subseteq U$.

Conversely, let $(S, \mathcal{T})$ is a T_1-space such that for any closed $A \subseteq S$ and every open set U containing A, there is an open set V containing A such that $\overline{V} \subseteq U$. Let A and B be disjoint closed sets in S. Then $U = S \setminus B$ is an open set containing A. So, by our assumption, there is an open set W containing A such that $\overline{W} \subseteq U$. Then W and $S \setminus \overline{W}$ are open sets, $A \subseteq W$, $B \subseteq S \setminus \overline{W}$ (if $x \in B$, then $x \notin S \setminus B = U$, and so, $x \notin \overline{W}$). Also, $W \cap (S \setminus \overline{W}) \subseteq W \cap (S \setminus W) = \emptyset$. □

LEVEL 3

5. Let $(S, \mathcal{T})$ be a T_2-space and $A \subseteq S$. Prove that $(A, \mathcal{T}_A)$ is a $\mathcal{T}_2$-space (Recall that $\mathcal{T}_A$ is the subspace topology on A). Determine if the analogous statement is true for T_3-spaces.

Proof: Let $x, y \in A$ with $x \neq y$. Since $(S, \mathcal{T})$ is a T_2-space, there are $U, V \in \mathcal{T}$ with $x \in U$, $y \in V$, and $U \cap V = \emptyset$. Since $x \in A$ and $x \in U$, $x \in A \cap U$. Since $y \in A$ and $y \in V$, $y \in A \cap V$. By the definition of $\mathcal{T}_A$, $A \cap U$ and $A \cap V$ are in $\mathcal{T}_A$. Finally, $(A \cap U) \cap (A \cap V) = A \cap (U \cap V) = A \cap \emptyset = \emptyset$. Since $x, y \in A$ were arbitrary, $(A, \mathcal{T}_A)$ is a T_2-space.

Let $(S, \mathcal{T})$ be a T_3-space and $A \subseteq S$. We will show that $(A, \mathcal{T}_A)$ is a $\mathcal{T}_3$-space.

Since $(S, \mathcal{T})$ is a T_3-space, $(S, \mathcal{T})$ is also a T_2-space. By the first paragraph, $(A, \mathcal{T}_A)$ is a T_2-space. Since every T_2-space is a T_1-space, $(A, \mathcal{T}_A)$ is a T_1-space.

Let $x \in A$ and $B \subseteq A \setminus \{x\}$ with B closed in $\mathcal{T}_A$. By definition, $A \setminus B$ is open in $\mathcal{T}_A$. So, there is $C \in \mathcal{T}$ with $A \setminus B = A \cap C$. Since $C \in \mathcal{T}$, $S \setminus C$ is closed in $(S, \mathcal{T})$. Since $x \in A \setminus B$, $x \in A \cap C$. So, $x \notin S \setminus C$. Since $(S, \mathcal{T})$ is a T_3-space, there are open sets $U, V \in \mathcal{T}$ with $x \in U, S \setminus C \subseteq V$, and $U \cap V = \emptyset$. Since $x \in A$ and $x \in U$, $x \in A \cap U$. Let $b \in B$. Then since $B \subseteq A \setminus \{x\}$, $b \in A$. Since $b \notin A \setminus B$, $b \notin A \cap C$. Since $b \in A$, $b \notin C$. So, $b \in S \setminus C$. Since $b \in B$ was arbitrary, $B \subseteq S \setminus C$. Since $S \setminus C \subseteq V$, $B \subseteq V$. Since $B \subseteq A$ and $B \subseteq V$, $B \subseteq A \cap V$. Finally, $(A \cap U) \cap (A \cap V) = A \cap (U \cap V) = A \cap \emptyset = \emptyset$. Since $x \in A$ was arbitrary and $B \subseteq A \setminus \{x\}$ was an arbitrary closed set, $(A, \mathcal{T}_A)$ is a T_3-space. □

6. Let $(S, \mathcal{T})$ be a topological space and let $D = \{(s, s) \mid s \in S\}$. Prove that $(S, \mathcal{T})$ is a T_2-space if and only if D is closed in $S \times S$ in the product topology.

Proof: First assume that $(S, \mathcal{T})$ is a T_2-space. We will prove that $(S \times S) \setminus D$ is open in $S \times S$. Let $(x, y) \in (S \times S) \setminus D$. Then $x \neq y$. Since $(S, \mathcal{T})$ is a T_2-space, there are open sets U and V with $x \in U$, $y \in V$, and $U \cap V \neq \emptyset$. So, $(x, y) \in U \times V$ and $(U \times V) \cap D = \emptyset$. Therefore, $U \times V \subseteq (S \times S) \setminus D$.

Conversely, assume that D is closed in $S \times S$ in the product topology. Let $x, y \in S$ with $x \neq y$. Then $(x, y) \notin D$, and so, (x, y) is in the open set $(S \times S) \setminus D$. Therefore, there is a basic open set $U \times V$ in the product topology such that $(x, y) \in U \times V$ and $U \times V \subseteq (S \times S) \setminus D$. Note that U and V are open in S, $x \in U$, and $y \in V$. If $z \in U \cap V$, then $(z, z) \in U \times V$. Since $U \times V \subseteq (S \times S) \setminus D$, $(z, z) \notin D$, which is clearly not true. So, $U \cap V = \emptyset$. □

LEVEL 4

7. Let $(S_1, \mathcal{T}_1)$ and $(S_2, \mathcal{T}_2)$ be T_2-spaces. Prove that $S_1 \times S_2$ with the product topology is also a T_2-space. Determine if the analogous statement is true for T_3-spaces.

Proof: Let $(x,y),(z,w) \in S_1 \times S_2$ with $(x,y) \neq (z,w)$. Then $x \neq z$ or $y \neq w$. Without loss of generality, assume that $x \neq z$. Since $(S_1, \mathcal{T}_1)$ is a T_2-space, there are $U, V \in \mathcal{T}_1$ with $x \in U$, $z \in V$, and $U \cap V = \emptyset$. Then $(x,y) \in U \times S_1$, $(z,w) \in V \times S_2$, and

$$(U \times S_1) \cap (V \times S_2) = (U \cap V) \times (S_1 \cap S_2) = \emptyset \times (S_1 \cap S_2) = \emptyset.$$

Since $(x,y),(z,w) \in S_1 \times S_2$ were arbitrary, $S_1 \times S_2$ with the product topology is a T_2-space.

Let $(S_1, \mathcal{T}_1)$ and $(S_2, \mathcal{T}_2)$ be T_3-spaces. We will show that $S_1 \times S_2$ with the product topology is also a T_3-space.

Let $(x,y) \in S_1 \times S_2$ and $B \subseteq (S_1 \times S_2) \setminus \{(x,y)\}$ with B closed in the product topology. Consider the open set $(S_1 \times S_2) \setminus B$. Since $(x,y) \in (S_1 \times S_2) \setminus B$, there are sets $U \in \mathcal{T}_1$ and $V \in \mathcal{T}_2$ with $x \in U$, $y \in V$, and $U \times V \subseteq (S_1 \times S_2) \setminus B$. Since U and V are open sets in $(S_1, \mathcal{T}_1)$ and $(S_2, \mathcal{T}_2)$, respectively, $S_1 \setminus U$ and $S_2 \setminus V$ are closed sets in $(S_1, \mathcal{T}_1)$ and $(S_2, \mathcal{T}_2)$, respectively. Also, $x \notin S_1 \setminus U$ and $y \notin S_2 \setminus V$. Since $(S_1, \mathcal{T}_1)$ and $(S_2, \mathcal{T}_2)$ are T_3-spaces, there are open sets W_1, Z_1, W_2, and Z_2 with $x \in W_1$, $y \in W_2$, $S_1 \setminus U \subseteq Z_1$, $S_2 \setminus V \subseteq Z_2$, $W_1 \cap Z_1 = \emptyset$, and $W_2 \cap Z_2 = \emptyset$.

Since $x \in W_1$ and $y \in W_2$, $(x,y) \in W_1 \times W_2$.

We now show that $B \subseteq (Z_1 \times S_2) \cup (S_1 \times Z_2)$. To see this, let $(a,b) \in B$. Then $(a,b) \notin U \times V$. So, $a \notin U$ or $b \notin V$. Without loss of generality, assume that $a \notin U$. Then $a \in Z_1$. So, $(a,b) \in Z_1 \times S_2$. It follows that $(a,b) \in (Z_1 \times S_2) \cup (S_1 \times Z_2)$.

Finally, we have

$$\begin{aligned}(W_1 \times W_2) \cap [(Z_1 \times S_2) \cup (S_1 \times Z_2)] &= [(W_1 \times W_2) \cap (Z_1 \times S_2)] \cup \big[(W_1 \times W_2) \cap \big((S_1 \times Z_2)\big)\big] \\ &= [(W_1 \cap Z_1) \times (W_2 \cap S_2)] \cup [(W_1 \cap S_1) \times (W_2 \cap Z_2)] \\ &= [\emptyset \times (W_2 \cap S_2)] \cup [(W_1 \cap S_1) \times \emptyset] = \emptyset \times \emptyset = \emptyset.\end{aligned}$$

Since $(x,y) \in S_1 \times S_2$ was arbitrary and $B \subseteq (S_1 \times S_2) \setminus \{(x,y)\}$ was an arbitrary closed set, $S_1 \times S_2$ with the product topology is a T_3-space. □

8. Prove that a second-countable T_3-space is a T_4-space.

Proof: Let $(S, \mathcal{T})$ be a T_3-space with a countable basis $\mathcal{B}$ and let A and B be disjoint closed sets in S. For each $x \in A$, $S \setminus B$ is an open set containing A. By Problem 3, there is $V_x \in \mathcal{B}$ containing x with $\overline{V_x} \subseteq S \setminus B$. The collection is countable, and so we can write $\{V_x \mid x \in S\} = \{V_k \mid k \in \mathbb{N}\}$. Note that we have $A \subseteq \bigcup\{V_n \mid n \in \mathbb{N}\}$ and for each $n \in \mathbb{N}$, $B \cap \overline{V}_n = \emptyset$.

In the same way, we can find $\{W_k \mid k \in \mathbb{N}\}$ with $B \subseteq \bigcup\{W_k \mid k \in \mathbb{N}\}$ and for each $n \in \mathbb{N}$, $A \cap \overline{W}_n = \emptyset$.

For each $n \in \mathbb{N}$, let $C_n = V_n \setminus \bigcup\{\overline{W_k} \mid k \in \mathbb{N}\}$ and $D_n = W_n \setminus \bigcup\{\overline{V_k} \mid k \in \mathbb{N}\}$. For each $n \in \mathbb{N}$, C_n and D_n are open in S. Let $V = \bigcup\{C_n \mid n \in \mathbb{N}\}$ and $W = \bigcup\{D_n \mid n \in \mathbb{N}\}$. Then V and W are open in S.

Since $A \subseteq \bigcup\{V_n \mid n \in \mathbb{N}\}$ and $A \cap \overline{W}_n = \emptyset$, we have $A \subseteq U$. Since $B \subseteq \bigcup\{W_n \mid n \in \mathbb{N}\}$ and $B \cap \overline{V}_n = \emptyset$, we have $B \subseteq V$.

If $x \in V \cap W$, then there are $m, n \in \mathbb{N}$ such that $x \in C_m \cap D_n$. Without loss of generality, assume that $m \geq n$. Then $x \in C_m = V_m \setminus \bigcup\{\overline{W_k} \mid k \in \mathbb{N}\} \subseteq V_m \setminus \overline{W_n}$ and $x \in D_n \subseteq W_n$, which is impossible. Therefore, $V \cap W = \emptyset$.

So, V and W are disjoint open sets with $A \subseteq V$ and $B \subseteq W$. □

LEVEL 5

9. Consider the topological space $(\mathbb{R}, \mathcal{T}_L)$ (see Problem 14 from Problem Set 9). Prove that $\mathbb{R}^2$ with the corresponding product topology is a T_3-space, but not a T_4-space.

Proof: We first show that $(\mathbb{R}, \mathcal{T}_L)$ is a T_4-space. Since $\mathcal{T}_L$ is finer than the standard topology on $\mathbb{R}$, and the standard topology is a T_1-space, $\mathcal{T}_L$ is also a T_1-space. Now, let A, B be disjoint closed subsets of $\mathbb{R}$. For each $a \in A$, we have $a \notin B$. So, $a \in \mathbb{R} \setminus B$. Since $\mathbb{R} \setminus B$ is open, there is a basic open set $[c, x_a)$ containing a such that $[c, x_a) \subseteq \mathbb{R} \setminus B$. Since $c \leq a$, we have $[a, x_a) \subseteq [c, x_a)$, and therefore, $[a, x_a) \subseteq \mathbb{R} \setminus B$. So, $[a, x_a) \cap B = \emptyset$. Similarly, for each $b \in B$, we can find x_b so that $[b, x_b) \cap A = \emptyset$. Let $U = \bigcup\{[a, x_a) \mid a \in A\}$ and let $V = \bigcup\{[b, x_b) \mid b \in B\}$. U and V are unions of basic open sets, thus open. Clearly, $A \subseteq U$ and $B \subseteq V$.

We show that $U \cap V = \emptyset$. Suppose toward contradiction that $z \in U \cap V$. Then there is $a \in A$ and $b \in B$ with $z \in [a, x_a)$ and $z \in [b, x_b)$. Without loss of generality, assume that $a < b$. Since $z \in [b, x_b)$, we have $b \leq z$. Since $z \in [a, x_a)$, we have $z < x_a$. So, $a < b \leq z < x_a$. It follows that $b \in [a, x_a)$, contradicting $[a, x_a) \cap B = \emptyset$. This contradiction shows that $U \cap V = \emptyset$. So, $(\mathbb{R}, \mathcal{T}_L)$ is a T_4-space.

Since $(\mathbb{R}, \mathcal{T}_L)$ is a T_4-space, it is also a T_3-space.

Let $\mathcal{T}$ be the product topology on $\mathbb{R}^2$ with respect to the topology $\mathcal{T}_L$. By the proof of Problem 7, $(\mathbb{R}^2, \mathcal{T})$ is a T_3-space.

We will now show that $(\mathbb{R}^2, \mathcal{T})$ is **not** a T_4-space.

Assume toward contradiction that $(\mathbb{R}^2, \mathcal{T})$ is a T_4-space. Let $D = \{(x, -x) \mid x \in \mathbb{R}\}$. D is a closed set in the standard product topology on $\mathbb{R}$ (as are all lines). Since $\mathcal{T}_L$ is finer than the standard topology on $\mathbb{R}$, D is also closed in $(\mathbb{R}^2, \mathcal{T})$.

Furthermore, $\mathcal{T}_D$ is the discrete topology on D. To see this, observe that the point $(x, -x)$ is equal to the intersection of D with the basic open set $[x, x+1) \times [-x, -x+1)$. Therefore, every singleton set $\{(x, -x)\}$ is open in $\mathcal{T}_D$. It follows that all subsets of D are both open and closed in $(D, \mathcal{T}_D)$.

If $A \subseteq D$, since A is closed in $(D, \mathcal{T}_D)$ and D is closed in $(\mathbb{R}^2, \mathcal{T})$, it follows that, A is closed in $(\mathbb{R}^2, \mathcal{T})$ (Why?). So, for any $A \subseteq D$ with $A \neq \emptyset$ and $A \neq D$, both A and $D \setminus A$ are closed in $(\mathbb{R}^2, \mathcal{T})$. Since we are assuming that $(\mathbb{R}^2, \mathcal{T})$ is a T_4-space, we can find disjoint $U_A, V_A \in \mathcal{T}$ with $A \subseteq U_A$ and $D \setminus A \subseteq V_A$.

Define $f: \mathcal{P}(D) \to \mathcal{P}(\mathbb{Q} \times \mathbb{Q})$ by $f(\emptyset) = \emptyset$, $f(D) = \mathbb{Q} \times \mathbb{Q}$, and $f(A) = (\mathbb{Q} \times \mathbb{Q}) \cap U_A$ for $A \neq \emptyset$ and $A \neq D$. We show that f is injective.

Let $A, B \in \mathcal{P}(D)$, both nonempty, both not equal to D or each other. Without loss of generality, assume there is $(x, -x) \in A \setminus B$. Then $(x, -x) \in D \setminus B$. Therefore, $(x, -x) \in U_A \cap V_B$. Since $U_A \cap V_B$ is open and nonempty, by the Density Theorem, there is $(a, -a) \in (U_A \cap V_B) \cap (\mathbb{Q} \times \mathbb{Q})$. Therefore, we have $(a, -a) \in (\mathbb{Q} \times \mathbb{Q}) \cap U_A = f(A)$ and $(a, -a) \notin (\mathbb{Q} \times \mathbb{Q}) \cap U_B = f(B)$. So, $f(A) \neq f(B)$.

Also, note that if $A \in \mathcal{P}(D)$, with $A \neq \emptyset$ and $A \neq D$, then $f(A) = (\mathbb{Q} \times \mathbb{Q}) \cap U_A$ is not empty because $\mathbb{Q} \times \mathbb{Q}$ has nonempty intersection with any open set, and $f(A) = (\mathbb{Q} \times \mathbb{Q}) \cap U_A$ is not $\mathbb{Q} \times \mathbb{Q}$ because $(\mathbb{Q} \times \mathbb{Q}) \cap V_A \neq \emptyset$. It follows that f is injective.

So, $\mathcal{P}(D) \preccurlyeq \mathcal{P}(\mathbb{Q} \times \mathbb{Q}) \sim \mathcal{P}(\mathbb{Q}) \sim \mathcal{P}(\mathbb{N}) \sim \mathbb{R} \sim D$, contradicting Cantor's Theorem. □

10. Let K be a nonempty set and let S_k be a T_2-space for each $k \in K$. Prove that $\prod S_k$ (with the product topology) is a T_2-space. Determine if the same result is true if we replace T_2 by each of T_3 and T_4.

Proof: Let K be a nonempty set, let S_k be a T_2-space for each $k \in K$, and let $\mathbf{x}, \mathbf{y} \in \prod S_k$ with $\mathbf{x} \neq \mathbf{y}$. Then there is $n \in K$ such that $x_n \neq y_n$. Since S_n is a T_2-space, there are open sets U_n and V_n in S_n such that $x_n \in U_n$, $y_n \in V_n$, and $U_n \cap V_n = \emptyset$. Then $\mathbf{x} \in \prod U_k$, where $U_k = S_k$ for each $k \neq n$, and $\mathbf{y} \in \prod V_k$, where $V_k = S_k$ for each $k \neq n$. We have

$$\prod U_k \cap \prod V_k = \prod (U_k \cap V_k) = \emptyset \text{ because } U_n \cap V_n = \emptyset.$$

Since $\mathbf{x}, \mathbf{y} \in \prod S_k$ were arbitrary, $\prod S_k$ with the product topology is a T_2-space.

Let K be a nonempty set, let S_k be a T_3-space for each $k \in K$. We will use Problem 3 to show that $\prod S_k$ with the product topology is also a T_3-space.

Let $\mathbf{x} \in \prod S_k$ and let $U = \prod U_k$ be a basic open set with $\mathbf{x} \in U$. If $U_k = S_k$, let $V_k = S_k$ as well. If $U_k \neq S_k$, by Problem 3, we can find an open set V_k with $x_k \in V_k$ and $\overline{V}_k \subseteq U_k$. Then $V = \prod V_k$ is a basic open set with $\mathbf{x} \in V$ and $\overline{V} \subseteq U$. Once again, by Problem 3, $\prod S_k$ is a T_3-space.

We can use the counterexample from Problem 9 to see that in general the result is **not** true for T_4-spaces. □

11. Let K be a nonempty set and let S_k be a nonempty topological space for each $k \in K$. Prove that if $\prod S_k$ (with the product topology) is a T_2-space, then so is S_k for each $k \in K$. Determine if the same result is true if we replace T_2 by each of T_3 and T_4.

Proof: Let K be a nonempty set, let S_k be a topological space for each $k \in K$, and suppose that $\prod S_k$ is a T_2-space. Let $k \in K$ and let $a, b \in S_n$ with $a \neq b$. For each $k \in K$ with $k \neq n$, choose $z_k \in S_k$. Define $\mathbf{x}, \mathbf{y} \in \prod S_k$ by $x_n = a$, $y_n = b$, and $x_k = z_k$, $y_k = z_k$ if $k \neq n$. Since $\prod S_k$ is a T_2-space, there are basic open sets $\prod U_k$ and $\prod V_k$ with $\mathbf{x} \in \prod U_k$, $\mathbf{y} \in \prod V_k$ and $\prod U_k \cap \prod V_k = \emptyset$. We then have $a = x_n \in U_n$, $b = y_n \in V_n$, and $U_n \cap V_n = \emptyset$. (If $c \in U_n \cap V_n$, then define $\mathbf{w}$ by $w_n = c$ and $w_k = z_k$ for $k \neq n$. Then we have $\mathbf{w} \in \prod U_k \cap \prod V_k$, a contradiction.)

Next, let K be a nonempty set, let S_k be a topological space for each $k \in K$, and suppose that $\prod S_k$ is a T_3-space. Let $k \in K$ and let $a \in S_n$ and let A be a closed set in S_n not containing a. For each $k \in K$ with $k \neq n$, choose $z_k \in S_k$. Define $\mathbf{x} \in \prod S_k$ by $x_n = a$ and $x_k = z_k$ if $k \neq n$. Define $C = \prod C_k$, where $C_n = A$ and $C_k = S_k$ for $k \neq n$. Then C is closed in $\prod S_k$ and C does not contain $\mathbf{x}$. Since $\prod S_k$ is a T_3-space, there are basic open sets $\prod U_k$ and $\prod V_k$ with $\mathbf{x} \in \prod U_k$, $C \subseteq \prod V_k$ and $\prod U_k \cap \prod V_k = \emptyset$. We then have $a = x_n \in U_n$, $A = C_n \subseteq V_n$, and $U_n \cap V_n = \emptyset$. (If $c \in U_n \cap V_n$, then define $\mathbf{w}$ by $w_n = c$ and $w_k = z_k$ for $k \neq n$. Then we have $\mathbf{w} \in \prod U_k \cap \prod V_k$, a contradiction.)

Finally, let K be a nonempty set, let S_k be a topological space for each $k \in K$, and suppose that $\prod S_k$ is a T_4-space. Let $k \in K$ and let A and B be a disjoint closed set in S_n. Define $C = \prod C_k$ and $D = \prod D_k$, where $C_n = A$, $D_n = B$, and $C_k = D_k = S_k$ for $k \neq n$. Then C and D are disjoint closed sets in $\prod S_k$. Since $\prod S_k$ is a T_4-space, there are basic open sets $\prod U_k$ and $\prod V_k$ with $C \subseteq \prod U_k$, $D \subseteq \prod V_k$ and $\prod U_k \cap \prod V_k = \emptyset$. We then have $A = C_n \subseteq U_n$, $B = D_n \subseteq V_n$, and $U_n \cap V_n = \emptyset$. (If $c \in U_n \cap V_n$, then define $\mathbf{w}$ by $w_n = c$ and for $k \neq n$, $w_k = z_k$ for some $z_k \in S_k$. Then we have $\mathbf{w} \in \prod U_k \cap \prod V_k$, a contradiction.) □

12. Prove that an uncountable product of first-countable topological spaces need not be first-countable.

Proof: Let K be an uncountable index set and for each $k \in K$, consider $S_k = \mathbb{Z}$ with the discrete topology. By part 1 of Example 10.13, each S_k is first-countable. We now show that $\prod S_k = {}^K\mathbb{Z}$ is not first-countable. To see this, let $\mathbf{x} \in {}^K\mathbb{Z}$ and let $\mathcal{D}$ be a countable family of basic open sets containing $\mathbf{x}$. Then for each $\prod B_k \in \mathcal{D}$, we have $B_k = S_k$ for all but finitely many $k \in \mathbb{Z}$. Therefore, there is $j \in K$ such that for every product $\prod B_k \in \mathcal{D}$, we have the jth coordinate set $B_j = S_j$. Let $C = \prod C_k$ be defined by $C_k = S_k$ for $k \neq j$ and $C_j = \{x_j\}$. Then C is an open set containing $\mathbf{x}$ that does not contain any open set in $\mathcal{D}$. It follows that $\mathcal{D}$ is not a countable basis at $\mathbf{x}$. Therefore, ${}^K\mathbb{Z}$ with the product topology is **not** first-countable. □

Problem Set 11

LEVEL 1

1. Let (S, d) be a metric space. Prove that for all $x, y \in S$, $d(x, y) \geq 0$.

Proof: Let (S, d) be a metric space and let $x, y \in S$. Then

$$2d(x, y) = d(x, y) + d(x, y) = d(x, y) + d(y, x) \geq d(x, x) = 0.$$

So, $d(x, y) \geq 0$, as desired. □

2. Prove that a sequence (s_n) is bounded in $\mathbb{R}$ (with the Euclidean metric) if and only if there is $M \in \mathbb{R}^+$ such that $|s_n| \leq M$ for all $n \in \mathbb{N}$.

Proof: Suppose that (s_n) is bounded in $\mathbb{R}$. Then there is $L \in \mathbb{R}^+$ such that for all $n, m \in \mathbb{N}$, $d(s_n, s_m) = |s_n - s_m| \leq L$. By the Triangle Inequality (and SACT), we have

$$|s_n| = |(s_n - s_0) + s_0| \leq |s_n - s_0| + |s_0| \leq L + |s_0|.$$

So, if we let $M = L + |s_0|$. Then for all $m \in \mathbb{N}$, $|x_m| \leq M$.

Conversely, suppose that there is $M \in \mathbb{R}^+$ such that $|s_n| \leq M$ for all $n \in \mathbb{N}$. Then for all $n, m \in \mathbb{N}$, we have

$$d(s_n, s_m) = |s_n - s_m| \leq |s_n| + |s_m| \leq 2M.$$

So (s_n) is bounded in $\mathbb{R}$ with the Euclidean metric. □

LEVEL 2

3. Let (S, d) be a metric space and let $\overline{d}$ be the standard bounded metric corresponding to d. Prove that $(S, \overline{d})$ is a metric space and that $\overline{d}$ induces the same topology as d.

Proof: Let's check that the 3 properties of a metric space are satisfied for $(S, \overline{d})$. We have $\overline{d}(x, y) = 0$ if and only if $\min\{1, d(x, y)\} = 0$ if and only if $d(x, y) = 0$ if and only if $x = y$. This shows that Property 1 holds. $d(x, y) = d(y, x)$ implies that $\overline{d}(x, y) = \min\{1, d(x, y)\} = \min\{1, d(y, x)\} = \overline{d}(y, x)$, and so, Property 2 holds. For Property 3, let $x, y, z \in S$. If $d(x, z) < 1$, then

$$\overline{d}(x, z) = d(x, z) \leq d(x, y) + d(y, z),$$

$$\overline{d}(x, z) = d(x, z) \leq 1 \leq d(x, y) + 1,$$

$$\overline{d}(x, z) = d(x, z) \leq 1 \leq 1 + d(y, z),$$

$$\overline{d}(x, z) = d(x, z) \leq 1 \leq 1 + 1,$$

Therefore, $\overline{d}(x, z) \leq \min\{1, d(x, y)\} + \min\{1, d(y, z)\} = \overline{d}(x, y) + \overline{d}(y, z)$.

Otherwise $d(x,z) \geq 1$. In this case, $\bar{d}(x,z) = 1$. If $d(x,y) > 1$ or $d(y,z) > 1$, then $\bar{d}(x,y) = 1$ or $\bar{d}(y,x) = 1$, and so, $\bar{d}(x,y) + \bar{d}(y,z) \geq 1$, and so, $\bar{d}(x,z) = 1 \leq \bar{d}(x,y) + \bar{d}(y,z)$.

Otherwise, we have $\bar{d}(x,z) = 1$, $d(x,y) \leq 1$, and $d(y,z) \leq 1$. So, $\bar{d}(x,y) = d(x,y)$ and $\bar{d}(y,z) = d(y,z)$. Suppose toward contradiction that $\bar{d}(x,z) > \bar{d}(x,y) + \bar{d}(y,z)$. Then we have $d(x,z) \geq 1 > d(x,y) + d(y,z)$, contradicting that (S,d) is a metric space.

Let $B_r(a;\bar{d})$ be an arbitrary open ball in $(S,\bar{d})$. We will show that $B_r(a;d) \subseteq B_r(a;\bar{d})$. To see thias, let $x \in B_r(a;d)$. Then $d(x,a) < r$. So, $\bar{d}(x,a) \leq d(x,a) < r$. Therefore, $x \in B_r(a;\bar{d})$. Since $x \in B_r(a;d)$ was arbitrary, it follows that $B_r(a;d) \subseteq B_r(a;\bar{d})$. So, the topology induced by d is finer than the topology induced by $\bar{d}$.

Now, let $B_r(a;d)$ be an arbitrary open ball in (S,d) and let $s = \min\{r,1\}$. We will show that $B_s(a;\bar{d}) \subseteq B_r(a;d)$. To see this, let $x \in B_s(a;\bar{d})$. Then $\bar{d}(x,a) < s$. Since $s \leq 1$, $\bar{d}(x,a) < 1$. It follows that $\bar{d}(x,a) = d(x,a)$. Therefore, $d(x,a) < s \leq r$. So, $x \in B_r(a;d)$. Since $x \in B_s(a;\bar{d})$ was arbitrary, it follows that $B_s(a;\bar{d}) \subseteq B_r(a;d)$. So, the topology induced by $\bar{d}$ is finer than the topology induced by d.

Since the topology induced by d is finer than the topology induced by $\bar{d}$ and the topology induced by $\bar{d}$ is finer than the topology induced by d, d and $\bar{d}$ induce the same topology. □

4. Let $d: \mathbb{R} \times \mathbb{R} \to \mathbb{R}$ be the Euclidean metric on $\mathbb{R}$, let $\bar{d}: \mathbb{R} \times \mathbb{R} \to \mathbb{R}$ be the standard bounded metric corresponding to d, and let $d^*: {}^{\mathbb{N}}\mathbb{R} \times {}^{\mathbb{N}}\mathbb{R} \to \mathbb{R}$ be the metric on $\mathbb{R}^n$ defined by $d^*(\mathbf{x},\mathbf{y}) = \sup\left\{\frac{\bar{d}(x_k,y_k)}{k+1} \,\middle|\, k \in \mathbb{N}\right\}$ (see Theorem 11.12). Prove each of the following:
   (i) (s_n) is a Cauchy sequence in $(\mathbb{R},d)$ if and only if (s_n) is a Cauchy sequence in $(\mathbb{R},\bar{d})$.
   (ii) If $(\mathbf{x}^n)$ is a Cauchy sequence in $({}^{\mathbb{N}}\mathbb{R},d^*)$, then for each $k \in K$, the sequence (x_k^n) is a Cauchy sequence in $(\mathbb{R},d)$.

Proofs:

(i) Suppose that (s_n) is a Cauchy sequence in $(\mathbb{R},d)$ and let $r \in \mathbb{R}^+$. Then there is $K \in \mathbb{N}$ such that $m \geq n > K$ implies $d(s_m,s_n) < r$. Since $\bar{d}(s_m,s_n) = \min\{1, d(s_m,s_n)\} \leq d(s_m,s_n)$, we have that $m \geq n > K$ implies $\bar{d}(s_m,s_n) < r$. It follows that (s_n) is a Cauchy sequence in $(\mathbb{R},\bar{d})$.

Conversely, suppose that (s_n) is a Cauchy sequence in $(\mathbb{R},\bar{d})$ and let $r \in \mathbb{R}^+$. Let $t = \min\{r,1\}$. Then there is $K \in \mathbb{N}$ such that $m \geq n > K$ implies $\bar{d}(s_m,s_n) < t$. Since $t \leq 1$, it follows that for $m \geq n > K$, $\bar{d}(s_m,s_n) = d(s_m,s_n)$. So, $m \geq n > K$ implies $d(s_m,s_n) < t \leq r$. It follows that (s_n) is a Cauchy sequence in $(\mathbb{R},d)$. □

(ii) Suppose that $(\mathbf{x}^n)$ is a Cauchy sequence in $({}^{\mathbb{N}}\mathbb{R}, d^*)$, let $k \in K$, let $r \in \mathbb{R}^+$, and let $t = \min\{r, 1\}$. Then there is $K \in \mathbb{N}$ such that $m \geq n > K$ implies $d^*(\mathbf{x}^m, \mathbf{x}^n) < \frac{t}{k}$. Therefore, $m \geq n > K$ implies $\sup\left\{\frac{\overline{d}(x_j^m, x_j^n)}{j} \,\middle|\, j \in \mathbb{N}\right\} < \frac{t}{k}$. So, $\frac{\overline{d}(x_k^m, x_k^n)}{k} \leq \sup\left\{\frac{\overline{d}(x_j^m, x_j^n)}{j} \,\middle|\, j \in \mathbb{N}\right\} < \frac{t}{k}$. Thus, $\overline{d}(x_k^m, y_k^n) < t$. Since $t \leq 1$, it follows that for $m \geq n > K$, $\overline{d}(x_k^m, x_k^n) = d(x_k^m, x_k^n)$. So, $m \geq n > K$ implies $d(x_k^m, x_k^n) < t \leq r$. It follows that (x_k^n) is a Cauchy sequence in $(\mathbb{R}, d)$. □

LEVEL 3

5. Define the functions d_1 and d_2 from $\mathbb{C} \times \mathbb{C}$ to $\mathbb{R}$ by $d_1(z, w) = |\text{Re } z - \text{Re } w| + |\text{Im } z - \text{Im } w|$ and $d_2(z, w) = \max\{|\text{Re } z - \text{Re } w|, |\text{Im } z - \text{Im } w|\}$. Prove that $(\mathbb{C}, d_1)$ and $(\mathbb{C}, d_2)$ are metric spaces such that d_1 and d_2 induce the standard topology on $\mathbb{C}$.

Proof: $d_1(z, w) = 0$ if and only if $|\text{Re } z - \text{Re } w| + |\text{Im } z - \text{Im } w| = 0$ if and only if $|\text{Re } z - \text{Re } w| = 0$ and $|\text{Im } z - \text{Im } w| = 0$ if and only if $\text{Re } z - \text{Re } w = 0$ and $\text{Im } z - \text{Im } w = 0$ if and only if $\text{Re } z = \text{Re } w$ and $\text{Im } z = \text{Im } w$ if and only if $z = w$. So, property 1 holds for d_1. Property 2 follows immediately from the fact that $|x - y| = |y - x|$ for all $x, y \in \mathbb{R}$. Let's verify property 3. Let $z, w, v \in \mathbb{C}$. Then, we have

$$d_1(z, v) = |\text{Re } z - \text{Re } v| + |\text{Im } z - \text{Im } v|$$

$$= |\text{Re } z - \text{Re } w + \text{Re } w - \text{Re } v| + |\text{Im } z - \text{Im } w + \text{Im } w - \text{Im } v| \text{ (by SACT)}$$

$$\leq |\text{Re } z - \text{Re } w| + |\text{Re } w - \text{Re } v| + |\text{Im } z - \text{Im } w| + |\text{Im } w - \text{Im } v| \text{ (by the Triangle Inequality)}$$

$$= |\text{Re } z - \text{Re } w| + |\text{Im } z - \text{Im } w| + |\text{Re } w - \text{Re } v| + |\text{Im } w - \text{Im } v| = d_1(z, w) + d_1(w, v).$$

This shows that $(\mathbb{C}, d_1)$ is a metric space.

$d_2(z, w) = 0$ if and only if $\max\{|\text{Re } z - \text{Re } w|, |\text{Im } z - \text{Im } w|\} = 0$ if and only if $|\text{Re } z - \text{Re } w| = 0$ and $|\text{Im } z - \text{Im } w| = 0$ if and only if $\text{Re } z - \text{Re } w = 0$ and $\text{Im } z - \text{Im } w = 0$ if and only if $\text{Re } z = \text{Re } w$ and $\text{Im } z = \text{Im } w$ if and only if $z = w$. So, property 1 holds for d_2. Property 2 follows immediately from the fact that $|x - y| = |y - x|$ for all $x, y \in \mathbb{R}$. Let's verify property 3. Let $z, w, v \in \mathbb{C}$. Then, we have

$$d_2(z, v) = \max\{|\text{Re } z - \text{Re } v|, |\text{Im } z - \text{Im } v|\}$$

$$= \max\{|\text{Re } z - \text{Re } w + \text{Re } w - \text{Re } v|, |\text{Im } z - \text{Im } w + \text{Im } w - \text{Im } v|\} \text{ (by SACT)}$$

$$\leq \max\{|\text{Re } z - \text{Re } w| + |\text{Re } w - \text{Re } v|, |\text{Im } z - \text{Im } w| + |\text{Im } w - \text{Im } v|\} \text{ (by the Triangle Inequality)}$$

$$\leq \max\{|\text{Re } z - \text{Re } w|, |\text{Im } z - \text{Im } w|\} + \max\{|\text{Re } w - \text{Re } v|, |\text{Im } w - \text{Im } v|\}$$

(In fact, it's not hard to show that for all $a, b, c, d \in \mathbb{R}$, $\max\{a + b, c + d\} \leq \max\{a, c\} + \max\{b, d\}$)

$$= d_2(z, w) + d_2(w, v).$$

This shows that $(\mathbb{C}, d_2)$ is a metric space.

Let $d: \mathbb{C} \times \mathbb{C} \to \mathbb{R}$ be defined by $d(z, w) = |z - w|$. We have already seen in part 1 of Example 14.10 that d induces the standard topology on $\mathbb{C}$. Let's let $\mathcal{T}$ be the standard topology on $\mathbb{C}$.

Now, if $z, w \in \mathbb{C}$, then we have

$$\max\{|\text{Re } z - \text{Re } w|, |\text{Im } z - \text{Im } w|\} \leq \sqrt{(\text{Re } z - \text{Re } w)^2 + (\text{Im } z - \text{Im } w)^2} = |z - w|$$
$$\leq |\text{Re } z - \text{Re } w| + |\text{Im } z - \text{Im } w| \leq 2\max\{|\text{Re } z - \text{Re } v|, |\text{Im } z - \text{Im } v|\}.$$

Therefore, $d_2(z,w) \leq d(z,w) \leq d_1(z,w) \leq 2d_2(z,w)$.

So, if $z \in \mathbb{C}$ and $r \in \mathbb{R}^+$, then $B_{\frac{r}{2}}(z; d_2) \subseteq B_r(z; d_1) \subseteq B_r(z; d) \subseteq B_r(z; d_2)$. For example, to see that $B_{\frac{r}{2}}(z; d_2) \subseteq B_r(z; d_1)$, if $w \in B_{\frac{r}{2}}(z; d_2)$, then $d_2(z,w) < \frac{r}{2}$, so that $2d_2(z,w) < r$. Then since $d_1(z,w) \leq 2d_2(z,w)$, $d_1(z,w) < r$, so that $w \in B_r(z; d_1)$. The other two arguments are similar.

Let U be an element of the topology induced by d_1. For each $z \in U$, let $B_{r_z}(z; d_1) \subseteq U$. Then for each $z \in U$, $B_{\frac{r_z}{2}}(z; d_2) \subseteq U$. Therefore, $\bigcup\left\{B_{\frac{r_z}{2}}(z; d_2)\middle|\, z \in U\right\} \subseteq U$. Also, if $w \in U$, then $w \in B_{\frac{r_w}{2}}(w; d_2)$, so that $w \in \bigcup\left\{B_{\frac{r_z}{2}}(z; d_2)\middle|\, z \in U\right\}$. Therefore, $U \subseteq \bigcup\left\{B_{\frac{r_z}{2}}(z; d_2)\middle|\, z \in U\right\}$. It follows that we have $U = \bigcup\left\{B_{\frac{r_z}{2}}(z; d_2)\middle|\, z \in U\right\}$. This shows that the topology $\mathcal{T}_2$ induced by d_2 is finer than the topology $\mathcal{T}_1$ induced by d_1. That is $\mathcal{T}_1 \subseteq \mathcal{T}_2$. Similarly, we have $\mathcal{T} \subseteq \mathcal{T}_1$ and $\mathcal{T}_2 \subseteq \mathcal{T}$. These inclusions together show us that $\mathcal{T} = \mathcal{T}_1 = \mathcal{T}_2$. So, d_1 and d_2 induce the standard topology on $\mathbb{C}$. □

6. Prove that the square metric on $\mathbb{R}^n$ induces the same topology as the Euclidean metric on $\mathbb{R}^n$. Then prove that the topology induced by these metrics is the same as the product topology on $\mathbb{R}^n$.

Proof: Let $B_r(\mathbf{x}; \rho)$ be an arbitrary open ball in $(\mathbb{R}^n, \rho)$. We will show that $B_{\frac{r}{\sqrt{n}}}(\mathbf{x}; \rho) \subseteq B_r(\mathbf{x}; d)$. To see this, let $\mathbf{y} \in B_{\frac{r}{\sqrt{n}}}(\mathbf{x}; \rho)$. Then $\max\{|y_1 - x_1|, \dots, |y_n - x_n|\} = \rho(\mathbf{y}, \mathbf{x}) < \frac{r}{\sqrt{n}}$. So, for each $k = 1, 2, \dots, n$, $|y_k - x_k| < \frac{r}{\sqrt{n}}$. Thus, $d(\mathbf{y}, \mathbf{x}) = \sqrt{(y_1 - x_1)^2 + (y_2 - x_2)^2 + (y_n - x_n)^2} \leq \sqrt{n(\frac{r}{\sqrt{n}})^2} = \sqrt{r^2} = r$. Therefore, $\mathbf{y} \in B_r(\mathbf{x}; d)$. So, the topology induced by ρ is finer than the topology induced by d.

Now, let $B_r(\mathbf{x}; d)$ be an arbitrary open ball in $(\mathbb{R}^n, d)$. We will show that $B_r(\mathbf{x}; d) \subseteq B_r(\mathbf{x}; \rho)$. To see this, let $\mathbf{y} \in B_r(\mathbf{x}; d)$. Then $\sqrt{(y_1 - x_1)^2 + (y_2 - x_2)^2 + (y_n - x_n)^2} = |\mathbf{y} - \mathbf{x}| = d(\mathbf{y}, \mathbf{x}) < r$. For each $k = 1, 2, \dots, n$, we have $|y_k - x_k| \leq \sqrt{(y_1 - x_1)^2 + (y_2 - x_2)^2 + (y_n - x_n)^2} = d(\mathbf{y}, \boldsymbol{x})$. Therefore, $\rho(\mathbf{y}, \mathbf{x}) = \max\{|y_1 - x_1|, \dots, |y_n - x_n|\} \leq d(\mathbf{y}, \mathbf{x}) < r$. So, $\mathbf{y} \in B_r(\mathbf{x}; \rho)$. Thus, the topology induced by d is finer than the topology induced by ρ.

Since the topology induced by ρ is finer than the topology induced by d and the topology induced by d is finer than the topology induced by ρ, ρ and d induce the same topology.

Now, let U be a basic open set in the product topology on $\mathbb{R}^n$, say $U = B_{r_1}(x_1) \times \cdots \times B_{r_n}(x_n)$, where for each $k = 1, 2, \dots, n$, $B_{r_k}(x_k)$ is an open ball in the Euclidean metric on $\mathbb{R}$. Let $r = \min\{r_1, \dots, r_n\}$. We will show that $B_r(\mathbf{x}; \rho) \subseteq U$. To see this, let $\mathbf{y} \in B_r(\mathbf{x}; \rho)$. Then

$$\max\{|y_1 - x_1|, \dots, |y_n - x_n|\} = \rho(\mathbf{y}, \mathbf{x}) < r.$$

So, for each $k = 1, 2, \dots, n$, we have $|y_k - x_k| < r \leq r_k$. So, for each $k = 1, 2, \dots, n$, $y_k \in B_{r_k}(x_k)$. It follows that $\mathbf{y} \in U$. Thus, the topology induced by ρ is finer that the product topology on $\mathbb{R}^n$.

Finally, let $B_r(\mathbf{x};\rho)$ be an arbitrary open ball in $(\mathbb{R}^n,\rho)$. We will show $B_r(x_1)\times\cdots\times B_r(x_n)\subseteq B_r(\mathbf{x};\rho)$. To see this, let $\mathbf{y}\in B_r(x_1)\times\cdots\times B_r(x_n)$. Then for each $k=1,2,\dots,n$, we have $y_k\in B_r(x_k)$. So, for each $k=1,2,\dots,n$, we have $|y_k-x_k|<r$. Therefore, $\rho(\mathbf{y},\mathbf{x})=\max\{|y_1-x_1|,\dots,|y_n-x_n|\}<r$. So, $\mathbf{y}\in B_r(\mathbf{x};\rho)$. Thus, the product topology on $\mathbb{R}^n$ is finer than the topology induced by ρ.

Since the topology induced by ρ is finer than the product topology on $\mathbb{R}^n$ and the product topology on $\mathbb{R}^n$ is finer than the topology induced by ρ, ρ induces the product topology on $\mathbb{R}^n$. □

LEVEL 4

7. Prove that every metrizable space is a T_4-space.

Proof: Let $(S,\mathcal{T})$ be metrizable and let d be a metric on S that induces $\mathcal{T}$. Let A,B be disjoint closed subsets of S. Let $x\in A$. Since $A\cap B=\emptyset$, $x\notin B$. So, x is in the open set $S\setminus B$. Therefore, there is $r_x\in\mathbb{R}^+$ such that $B_{r_x}(x)\subseteq S\setminus B$. Let $U=\bigcup\left\{B_{\frac{r_x}{2}}(x)\;\middle|\;x\in A\right\}$. Then $U\in\mathcal{T}$, $A\subseteq U$, and $U\cap B=\emptyset$. Similarly, for each $x\in B$, let $V=\bigcup\left\{B_{\frac{r_x}{2}}(x)\;\middle|\;x\in B\right\}$, so that $V\in\mathcal{T}$, $B\subseteq V$, and $V\cap A=\emptyset$.

We now show that $U\cap V=\emptyset$. If $a\in U\cap V$, then there is $x\in A$ and $y\in B$ with $a\in B_{\frac{r_x}{2}}(x)\cap B_{\frac{r_y}{2}}(y)$. Then $d(x,y)\le d(x,a)+d(a,y)<\frac{r_x}{2}+\frac{r_y}{2}$. Without loss of generality, assume that $r_y\le r_x$. Then, we have $d(x,y)<\frac{r_x}{2}+\frac{r_x}{2}=r_x$. So, $y\in B_{r_x}(x)$. Since $B_{r_x}(x)\subseteq S\setminus B$, $y\in S\setminus B$. So, $y\notin B$, a contradiction. It follows that $U\cap V=\emptyset$. □

8. Let $(S_1,\mathcal{T}_1)$ and $(S_2,\mathcal{T}_2)$ be metrizable spaces. Prove that $S_1\times S_2$ with the product topology is metrizable. Use this to show that $(\mathbb{R},\mathcal{T}_L)$ is not metrizable (see Problem 14 from Problem Set 9 for the definition of $\mathcal{T}_L$).

Proof: Let d_1 and d_2 be metrics that induce the topologies $\mathcal{T}_1$ and $\mathcal{T}_2$, respectively. Define $d\colon(S_1\times S_2)\times(S_1\times S_2)\to\mathbb{R}$ by $d\big((a,b),(c,d)\big)=\max\{d_1(a,c),d_2(b,d)\}$. We first show that d defines a metric on $S_1\times S_2$. We have $d\big((a,b),(c,d)\big)=0$ if and only if $\max\{d_1(a,c),d_2(b,d)\}=0$ if and only if $d_1(a,c)=0$ and $d_2(b,d)=0$ if and only if $a=c$ and $b=d$ if and only if $(a,b)=(c,d)$. So, property 1 holds. Property 2 is clear. For property 3, Let $(a,b),(c,k),(e,f)\in S_1\times S_2$, Then

$$d\big((a,b),(e,f)\big)=\max\{d_1(a,e),d_2(b,f)\}\le\max\{d_1(a,c)+d_1(c,e),d_2(b,k)+d_2(k,f)\}$$
$$\le\max\{d_1(a,c),d_2(b,k)\}+\max\{d_1(c,e),d_2(k,f)\}=d\big((a,b),(c,k)\big)+d\big((c,k),(e,f)\big).$$

We now show that d induces the product topology on $S_1\times S_2$.

Let $B=B_r\big((x,y);d\big)$ be an arbitrary open ball in the topology induced by d. We show that $B=B_r(x;d_1)\times B_r(y;d_2)$. If $(a,b)\in B$, then $\max\{d_1(x,a),d_2(y,b)\}=d\big((x,y),(a,b)\big)<r$. So, $d_1(x,a)<r$ and $d_2(y,b)<r$. Thus, $a\in B_r(x;d_1)$, $b\in B_r(y;d_2)$. So, $(a,b)\in B_r(x;d_1)\times B_r(y;d_2)$. Therefore, $B\subseteq B_r(x;d_1)\times B_r(y,d_2)$.

Now, if $(a,b) \in B_r(x;d_1) \times B_r(y;d_2)$, then $a \in B_r(x;d_1)$ and $b \in B_r(y;d_2)$. So, $d_1(x,a) < r$ and $d_2(y,b) < r$. So, $d((x,y),(a,b)) = \max\{d_1(x,a), d_2(y,b)\} < r$. Therefore, $(a,b) \in B$. So, $B_r(x;d_1) \times B_r(y;d_2) \subseteq B$. Since $B \subseteq B_r(x;d_1) \times B_r(y,d_2)$ and $B_r(x;d_1) \times B_r(y;d_2) \subseteq B$, we have $B \subseteq B_r(x;d_1) \times B_r(y,d_2)$. This shows that B is open in the product topology. Therefore, the product topology is finer than the topology induced by d.

Conversely, a basic open set in the product topology has the form $B_r(x;d_1) \times B_r(y;d_2)$, and we saw that this equal to $B_r((x,y);d)$. It follows that each basic open set in the product topology is open in the topology induced by d. Therefore, the topology induced by d is finer than the product topology. So, d induces the product topology.

Now, assume towards contradiction that $(\mathbb{R}, \mathcal{T}_L)$ is metrizable. It follows that $\mathbb{R}^2$ with the corresponding product topology is metrizable. By Problem 7, $\mathbb{R}^2$ with the product topology is a T_4-space, contradicting what we proved above. So, $(\mathbb{R}, \mathcal{T}_L)$ is **not** metrizable. □

LEVEL 5

9. Prove that $(\mathbb{R}, d)$, where d is the Euclidean metric, has the Bolzano-Weierstrass Property. You may use the following fact: if $\mathcal{C} = \{[a_k, b_k] \mid k \in \mathbb{N}\}$ is a sequence of closed intervals in $\mathbb{R}$, such that $j < k \to [a_k, b_k] \subseteq [a_j, b_j]$, then $\bigcap \mathcal{C} \neq \emptyset$ (Example 12.3 together with the Heine-Borel Theorem (Theorem 12.6) will explain why this is true).

Proof: Let (s_n) be a bounded sequence in $\mathbb{R}$. Since (s_n) is bounded, there are $a, b \in \mathbb{R}$ such that for all $n \in \mathbb{N}$, $a \leq s_n \leq b$, or equivalently, for all $n \in \mathbb{N}$, $s_n \in [a,b]$.

Let $a_0 = a$, $b_0 = b$, and $c_0 = \frac{a_0+b_0}{2}$. Either $[a_0, c_0]$ or $[c_0, b_0]$ contains infinitely many terms of the sequence. If $[a_0, c_0]$ contains infinitely many terms of the sequence, let $a_1 = a_0$, $b_1 = c_0$. Otherwise, let $a_1 = c_0$ and $b_1 = b_0$. Then let $c_1 = \frac{a_1+b_1}{2}$. Continuing in this way, we see that at stage n, we have that$[a_n, c_n]$ or $[c_n, b_n]$ contain infinitely many terms of the sequence. If $[a_n, c_n]$ contains infinitely many terms of the sequence, let $a_{n+1} = a_n$, $b_{n+1} = c_n$. Otherwise, let $a_{n+1} = c_n$ and $b_{n+1} = b_n$. Then let $c_{n+1} = \frac{a_n+b_n}{2}$. This gives us a sequence of nested intervals $[a_0, b_0] \supseteq [a_1, b_1] \supseteq \cdots \supseteq [a_n, b_n] \supseteq \cdots$, where each interval $[a_n, b_n]$ contains infinitely many terms of the given sequence and the length of $[a_n, b_n]$ is $\frac{b-a}{2^n}$. It follows that $\bigcap\{[a_n, b_n] \mid n \in \mathbb{N}\}$ contains at most one real number. By the fact given in the problem, we have $\bigcap\{[a_n, b_n] \mid n \in \mathbb{N}\} \neq \emptyset$. Therefore, $\bigcap\{[a_n, b_n] \mid n \in \mathbb{N}\}$ contains exactly one real number x.

We now construct a subsequence of (s_n) converging to x. Let $s_{n_0} = a_0$. Given $\left(s_{n_j}\right)_{j \leq k}$ with $n_0 < n_1 < \cdots$ and $s_{n_j} \in [a_j, b_j]$ for each $j \leq k$, we can choose $s_{n_{k+1}} \in [a_{k+1}, b_{k+1}]$ with $n_{k+1} > n_k$ (because $\left[a_{n_{k+1}}, b_{n_{k+1}}\right]$ contains infinitely many terms of the sequence).

We claim that $s_{n_k} \to x$. To see this, let $k \in \mathbb{N}^+$. Since $\frac{b-a}{2^n} \to 0$, there is $K \in \mathbb{N}$ such that $n > K$ implies $\frac{b-a}{2^n} = \left|\frac{b-a}{2^n}\right| < \frac{1}{k}$. Since $s_{n_k} \in [a_k, b_k]$ and $x \in [a_k, b_k]$, we have that $|s_{n_k} - x|$ is less than or equal to the length of $[a_k, b_k]$, and so, $|s_{n_k} - x| \leq \frac{b-a}{2^{n_k}} < \frac{1}{k}$. It follows that $s_{n_k} \to x$. □

10. Let (S, d) be a metric space. Prove that for any index set K, the uniform metric on ${}^K S$ is finer than the product topology on ${}^K S$.

Proof: Let $\mathbf{x} \in {}^K S$ and let $\prod U_k$ be a basic open set in the product topology of ${}^K S$. So, $U_k = S$ for all but finitely $k \in K$, say $U_k = S$ for all $k \in K \setminus L$, where L is a finite set.

For each $j \in L$, there is $r_j \in \mathbb{R}^+$ with $r_j < 1$ such that $B_{r_j}(x_j; d) \subseteq U_j$. Let $r = \min\{r_j \mid j \in L\}$. Suppose $\rho(\mathbf{x}, \mathbf{y}) < r$. Then for each $j \in L$, $d(x_j, y_j) = \overline{d}(x_j, y_j) \leq \sup\{\overline{d}(x_k, y_k) \mid k \in K\} = \rho(\mathbf{x}, \mathbf{y}) < r$. So, for all $j \in L$, $y_j \in B_{x_j}(r; d)$. Since $B_{x_j}(r; d) \subseteq U_j$, we have $y_j \in U_j$. If $j \in K \setminus L$, then $U_j = S$, and so, we trivially have that $y_j \in U_j$. So, if $\mathbf{y} \in B_r(\mathbf{x}; \rho)$, then $\mathbf{y} \in U$. Thus, $B_r(\mathbf{x}; \rho) \subseteq U$. Therefore, the uniform metric on ${}^K S$ is finer than the product topology on ${}^K S$. □

11. Prove that for any index set K, $({}^K\mathbb{R}, \rho)$ is a complete metric space, where ρ is the uniform metric on ${}^K\mathbb{R}$.

Proof: By Theorem 11.11, $(\mathbb{R}, d)$ is a complete metric space. By part (i) of Problem 4 above, the Cauchy sequences in $(\mathbb{R}, d)$ and $(\mathbb{R}, \overline{d})$ are the same. It follows that $(\mathbb{R}, \overline{d})$ is also a complete metric space.

Now, let $(\mathbf{x}^n)$ be a Cauchy sequence in $({}^K\mathbb{R}, \rho)$, let $k \in K$ and let $r \in \mathbb{R}^+$. Then there is $L \in \mathbb{N}$ such that $m \geq n > L$ implies $\rho(\mathbf{x}^m, \mathbf{x}^n) < r$. Therefore, $m \geq n > L$ implies $\sup\{\overline{d}(x_j^m, x_j^n) \mid j \in \mathbb{N}\} < r$. So, for each $k \in K$, $\overline{d}(x_k^m, x_k^n) \leq \sup\{\overline{d}(x_j^m, x_j^n) \mid j \in \mathbb{N}\} = \rho(\mathbf{x}^m, \mathbf{x}^n) < r$. It follows that for each $k \in K$, (x_k^n) is a Cauchy sequence in $(\mathbb{R}, \overline{d})$. Since $(\mathbb{R}, \overline{d})$ is complete, for each $k \in K$, there is x_k such that $x_k^n \to x_k$ in $(\mathbb{R}, \overline{d})$. Let $\mathbf{x} = (x_k) \in {}^K\mathbb{R}$.

We now show that $\mathbf{x}^n \to \mathbf{x}$. To that end, let $r \in \mathbb{R}^+$. Since $x_k^n \to x_k$ in $(\mathbb{R}, \overline{d})$, there is $L_1 \in \mathbb{N}$ such that $n > L_1$ implies that $\overline{d}(x_k^n, x_k) < \frac{r}{2}$. Since $(\mathbf{x}^n)$ is a Cauchy sequence in $(\mathbf{x}^n)$, there is $L_1 \in \mathbb{N}$ such that $m \geq n > L_2$ implies $\rho(\mathbf{x}^n, \mathbf{x}^m) < \frac{r}{2}$. Let $L = \max\{L_1, L_2\}$.

Let $m \geq n > L$ and let $k \in K$. Then we have

$$\begin{aligned}\overline{d}(x_k^n, x_k) &\leq \overline{d}(x_k^n, x_k^m) + \overline{d}(x_k^m, x_k) \leq \sup\{\overline{d}(x_j^n, x_j^m) \mid j \in \mathbb{N}\} + \overline{d}(x_k^m, x_k) \\ &= \rho(\mathbf{x}^n, \mathbf{x}^m) + \overline{d}(x_k^m, x_k) < \frac{r}{2} + \frac{r}{2} = r\end{aligned}$$

Since the left -hand side does not depend on m, we get that for $n > L$ and $k \in K$, $\overline{d}(x_k^n, x_k) \leq r$. So, for $n > L$, we have $\rho(\mathbf{x}^n, \mathbf{x}) \leq r$. Therefore, $\mathbf{x}^n \to \mathbf{x}$ in the uniform metric on ${}^K\mathbb{R}$. It follows that $({}^K\mathbb{R}, \rho)$ is a complete metric space. □

Problem Set 12

LEVEL 1

1. Let I be an open or half-open interval of real numbers. Prove that $(I, \mathcal{T}_I)$ is **not** compact.

Proof: Let $I = (a, b)$. The collection of open sets $\mathcal{C} = \left\{\left(a + \frac{1}{n}, b\right) \middle| n \in \mathbb{Z}^+\right\}$ is an open covering of I with no finite subcover.

Let $J = (a, b]$. The collection $\mathcal{C} = \left\{\left(a + \frac{1}{n}, b\right] \middle| n \in \mathbb{Z}^+\right\}$ is an open covering of J with no finite subcover.

Let $K = [a, b)$. The collection $\mathcal{C} = \left\{\left[a, b - \frac{1}{n}\right) \middle| n \in \mathbb{Z}^+\right\}$ is an open covering of K with no finite subcover. □

2. Let $(S, \mathcal{T})$ be a topological space. Prove that a finite union of compact subspaces of S is compact.

Proof: Let $(S, \mathcal{T})$ be a topological space, and for each $k = 0, 1, \dots, n$, let $\left(A_k, \mathcal{T}_{A_k}\right)$ be a compact subspace of $(S, \mathcal{T})$. Let $A = \bigcup\{A_k \mid k = 0, 1, \dots, n\}$. We will show that $(A, \mathcal{T}_A)$ is compact. To this end, let $\mathcal{C}$ be an open covering of A Then for each $k = 0, 1, \dots, n$, $\mathcal{C}$ is an open covering of A_k by open sets in S. By Lemma 12.10, for each $k = 0, 1, \dots, n$, there is a finite subcollection $\mathcal{C}_k$ such that $\mathcal{C}_k$ still covers A_k. Then $\bigcup\{\mathcal{C}_k \mid k = 0, 1, \dots, n\}$ is an finite subcollection of $\mathcal{C}$ that covers A. By Lemma 12.10 again, $(A, \mathcal{T}_A)$ is compact. □

3. Prove that the Cantor set is compact and nowhere dense in $\mathbb{R}$.

Proof: Let C be the Cantor set. Then C is a closed set in the compact space $[0, 1]$. By Theorem 12.13, $(C, \mathcal{T}_C)$ is compact.

Since the Cantor set is closed, $\overline{C} = C$, and so we need only show that C has empty interior. Suppose toward contradiction that $U \subseteq C$ is an open interval. Suppose that $\operatorname{diam} U = d$. By the Archimedean Property of $\mathbb{R}$, we can find $n \in \mathbb{N}$ with $n > \frac{1}{d}$, or equivalently, $\frac{1}{n} < d$. Now, $C = \bigcap C_k$, where C_k is the union of 2^k closed intervals, each of diameter $\frac{1}{3^k}$. In particular, C_n is the union of 2^n closed intervals, each of diameter $\frac{1}{3^n}$. Since $3^n > n$, we have $\frac{1}{3^n} < \frac{1}{n} < d$. Therefore, U cannot be contained inside of one of these intervals. It follows that there is $x \in U$ with $x \notin C_n$. Since $C \subseteq C_n$, $x \notin C$. Therefore, U is **not** a subset of C, contrary to our assumption. It follows that C has empty interior. □

LEVEL 2

4. Let $(A, \mathcal{T}_A)$ be a compact subspace of a T_1-space $(S, \mathcal{T})$. Is A necessarily closed in S? (Compare this problem with Theorem 12.12.)

Solution: No. Consider $(\mathbb{R}, \mathcal{T})$, where $\mathcal{T}$ is the cofinite topology on $\mathbb{R}$. By part 3 of Example 10.2, this is a T_1-space. If $A \subseteq \mathbb{R}$ is any subset, then $(A, \mathcal{T}_A)$ is compact. To see this, let $\mathcal{C}$ be an open covering of A by open sets in $\mathbb{R}$ and let A_0 be any set in $\mathcal{C}$. Then $S \setminus A_0$ is finite, say $S \setminus A_0 = \{a_1, a_2, \dots, a_n\}$. For each $i = 1, 2, \dots, n$, let $A_i \in \mathcal{C}$ with $a_i \in A_i$. Then the collection $\{A_0, A_1, A_2, \dots, A_n\}$ is a finite subcollection from $\mathcal{C}$ that covers S. By Lemma 12.10, $(A, \mathcal{T}_A)$ is compact.

Not every subset of $\mathbb{R}$ is closed in this topology. In fact, aside from $\mathbb{R}$ itself, the only closed sets in $\mathbb{R}$ are the finite sets. As a specific example, $(\mathbb{N}, \mathcal{T}_{\mathbb{N}})$ is compact by the previous paragraph, but it is not closed because $\mathbb{N}$ is infinite.

5. Prove that the following are equivalent for a topological space $(S, \mathcal{T})$:
   (i) $(S, \mathcal{T})$ is compact.
   (ii) Every covering of S by basic open sets contains a finite subcollection that still covers S.
   (iii) For every collection $\mathcal{C}$ of basic closed sets in S with the finite intersection property, we have $\bigcap \mathcal{C} \neq \emptyset$.

Proof: (i → ii) Suppose that $(S, \mathcal{T})$ is compact and let $\mathcal{C}$ be a covering of S by basic open sets. Then $\mathcal{C}$ is a covering of S by open sets. Since $(S, \mathcal{T})$ is compact, there is a finite subcollection of $\mathcal{C}$ that still cover S.

(ii → iii) Let $(S, \mathcal{T})$ be a topological space and suppose that there is a collection $\mathcal{C}$ of basic closed sets in S with the finite intersection property such that $\bigcap \mathcal{C} = \emptyset$. Let $\mathcal{D} = \{S \setminus B \mid B \in \mathcal{C}\}$. Then $\mathcal{D}$ is a collection of basic open sets in S. Since $\bigcap \mathcal{C} = \emptyset$, we have $\bigcup \mathcal{D} = S \setminus \bigcap \mathcal{C} = S \setminus \emptyset = S$ (by De Morgan's law). Let $\mathcal{E}$ be a finite subcollection of elements in $\mathcal{D}$ that covers S, say $\mathcal{E} = \{S \setminus B_1, S \setminus B_2, \dots, S \setminus B_k\}$. Then $\{B_1, B_2, \dots, B_k\}$ is a finite subcollection of $\mathcal{C}$. Since $\mathcal{C}$ has the finite intersection property, there is $x \in B_1 \cap B_2 \cap \dots \cap B_k$. So, $x \notin S \setminus (B_1 \cap B_2 \cap \dots \cap B_k) = (S \setminus B_1) \cup (S \setminus B_2) \cup \dots \cup (S \setminus B_k)$. It follows that $\mathcal{E}$ does not cover S. Since $\mathcal{E}$ was an arbitrary finite subcollection of $\mathcal{D}$, we have found a basic open covering of S without a finite subcollection that covers S.

(iii → ii) Suppose there is a basic open covering $\mathcal{C}$ of S such that no finite subcollection of $\mathcal{C}$ covers S. Let $\mathcal{D} = \{S \setminus U \mid U \in \mathcal{C}\}$. Then $\mathcal{D}$ is a collection of basic closed sets. Since $\bigcup \mathcal{C} = S$, by De Morgan's law, we have $\bigcap \mathcal{D} = S \setminus \bigcup \mathcal{C} = S \setminus S = \emptyset$. Let $\mathcal{E}$ be a finite subcollection of elements in $\mathcal{D}$, say $\mathcal{E} = \{S \setminus U_1, S \setminus U_2, \dots, S \setminus U_k\}$. Then $\{U_1, U_2, \dots, U_k\}$ is a finite subcollection of $\mathcal{C}$. Since $\mathcal{C}$ does not have a finite subcollection that covers S, there is $x \in S$ such that $x \notin U_1 \cup U_2 \cup \dots \cup U_k$. So, $x \in S \setminus (U_1 \cup U_2 \cup \dots \cup U_k) = (S \setminus U_1) \cap (S \setminus U_2) \cap \dots \cap (S \setminus U_k)$. Since $\mathcal{E}$ was an arbitrary finite subcollection of $\mathcal{D}$, we see that $\mathcal{D}$ has the finite intersection property. And yet, we saw that $\bigcap \mathcal{D} = \emptyset$.

(ii → i) Let $(S, \mathcal{T})$ be a topological space such that every covering of S by basic open sets contains a fnite subcollection that still covers S. Let $\mathcal{C}$ be an open covering of S. For each $x \in S$ and $U \in C$, let U_x be a basic open set containing x with $U_x \subseteq U$. Then $\mathcal{D} = \{U_x \mid x \in S \wedge U \in \mathcal{C}\}$ is a covering of S by basic open sets. By (ii), there is a finite subcollection $\mathcal{E}$ of $\mathcal{D}$ that still covers S. Let $\mathcal{F} = \{U \mid U_x \in \mathcal{E}\}$. Then $\mathcal{F}$ is a finite subcollection of $\mathcal{C}$ that covers S. Since $\mathcal{C}$ was an arbitrary open covering of S, $(S, \mathcal{T})$ is compact. □

LEVEL 3

6. Prove that a totally bounded metric space is bounded.

Proof: Let (S, d) be a totally bounded metric space and let A be a 1-net. Since A is finite, $m = \max\{d(a, b) \mid a, b \in A\}$ is finite. Let $M = m + 2$ and let $x, y \in S$ be arbitrary. Since A is a 1-net, there are $a, b \in A$ with $d(x, a) < 1$ and $d(y, b) < 1$. Then

$$d(x, y) \leq d(x, a) + d(a, b) + d(b, y) < 1 + m + 1 = m + 2 = M.$$

Therefore, (S, d) is bounded by M. □

7. Give a direct proof that if $(S, \mathcal{T})$ is a sequentially compact metrizable space, then every infinite subset of S has an accumulation point.

Proof: Assume that $(S, \mathcal{T})$ is sequentially compact and let A be an infinite subset of S. Then there is a sequence (s_n) in A such that $n \neq m \to s_n \neq s_m$. By sequential compactness, (s_n) has a convergent subsequence (s_{n_k}), let's say $s_{n_k} \to x$. Assume toward contradiction that x is not an accumulation point of A. Then there is an open ball $B_r(x)$ with center x such that $B_r(x) \cap A = \{x\}$. Since $s_{n_k} \to x$, there is $K \in \mathbb{Z}^+$ such that $s_{n_k} \in B_r(x) \cap A = \{x\}$ foir all $k > K$. So, for $k > K$, $s_{n_k} = x$, contradicting our assumption that all members of the sequence (and thus the subsequence) are distinct. □

8. Prove that every sequentially compact metric space is totally bounded.

Proof: Assume that (S, d) is a sequentially compact metric space and let $r \in \mathbb{R}^+$. Let $s_0 \in S$ be arbitrary. If $B_r(s_0)$ is not an r-net, let $s_1 \in S \setminus B_r(s_0)$. We continue inductively. If $B_r(s_0) \cup \cdots \cup B_r(s_n)$ is not an r-net, let $s_n \in S \setminus (B_r(s_0) \cup \cdots \cup B_r(s_n))$. This process must stop in finitely many steps, or else the sequence (s_k) would have no convergent subsequence. □

9. Prove that a metric space is compact if and only if it is complete and totally bounded.

Proof: First assume that (S, d) is a compact metric space. By Theorem 12.20, (S, d) is sequentially compact. If (s_n) is a Cauchy sequence in S, then by sequential compactness, (s_n) has a convergent subsequence. By Corollary 11.8, (S, d) is complete. By Problem 8 above, (S, d) is totally bounded.

Conversely, assume that (S, d) is a complete, totally bounded metric space. Let (s_n) be a sequence in S. We use the fact that (S, d) is totally bounded to show that (s_n) has a Cauchy subsequence. For each $n \in \mathbb{N}$, let A_n be a $\frac{1}{n+1}$-net. Since A_0 is finite, there is $a_0 \in A_0$ such that infinitely many terms of (s_n) are in $B_1(a_0)$. Let $C_0 = \{n \in \mathbb{N} \mid s_n \in B_1(a_0)\}$. Since A_1 is finite, there is $a_1 \in A_1$ such that infinitely many of the elements of C_0 are in $B_{\frac{1}{2}}(a_1)$. Let $C_1 = \left\{n \in C_0 \,\middle|\, s_n \in B_{\frac{1}{2}}(a_1)\right\}$. Continuing in this way, assuming that C_k is infinite, since A_{k+1} is finite, there is $a_{k+1} \in A_{k+1}$ such that infinitely many of the elements of C_k are in $B_{\frac{1}{k+2}}(a_{k+1})$. Let $C_{k+1} = \left\{n \in C_k \,\middle|\, s_n \in B_{\frac{1}{k+1}}(a_{k+1})\right\}$. Let (n_k) be a strictly increasing sequence of natural numbers such that $n_k \in C_k$ for each $k \in \mathbb{N}$. It is easy to check that (s_{n_k}) is a Cauchy sequence. Now, since (S, d) is complete, this Cauchy sequence converges, giving us a convergent subsequence of (s_n). It follows that (S, d) is sequentially compact. By Theorem 12.20, (S, d) is compact. □

LEVEL 4

10. Let $(S,\mathcal{T})$ be a compact T_2-space and let K, L be subsets of S that are each compact in the subspace topology. Prove that there are disjoint open sets U and V such that $K \subseteq U$ and $L \subseteq V$. Use this result to prove that a compact T_2-space is a T_4-space.

Proof: Let $(S,\mathcal{T})$ be a compact T_2-space and let K, L be subsets of S that are each compact in the subspace topology.

Let $x \in K$. Since $(S,\mathcal{T})$ is a T_2-space, for each $y \in L$, we can find open sets U_y and V_y such that $x \in U_y$, $y \in V_y$, and $U_y \cap V_y = \emptyset$. The collection $\mathcal{C} = \{V_y \mid y \in L\}$ is an open cover of L by open sets in S. By Lemma 12.10, there is a finite subcollection $\mathcal{D} \subseteq \mathcal{C}$ that still covers L. Let $V = \bigcup \mathcal{D}$ and let $W = \bigcap\{U_y \mid V_y \in \mathcal{D}\}$. Then V and W are disjoint open sets such that $x \in W$ and $L \subseteq V$.

Now, for each $x \in K$, by the previous paragraph, we can find disjoint open sets W_x and V_x such that $x \in W_x$ and $L \subseteq V_x$. The collection $\mathcal{E} = \{W_x \mid x \in K\}$ is an open cover of K by open sets in S. By Lemma 12.10, there is a finite subcollection $\mathcal{F} \subseteq \mathcal{E}$ that still covers K. Let $W = \bigcup \mathcal{F}$ and let $X = \bigcap\{V_x \mid W_x \in \mathcal{F}\}$. Then W and X are disjoint open sets such that $K \subseteq W$ and $L \subseteq V$. □

11. Let (A, d) be a subspace of $(\mathbb{R}^n, d)$, where d is the Euclidean metric. Prove that A is bounded if and only if A is totally bounded.

Proof: By Problem 6, if A is totally bounded, then A is bounded.

Now, let (A, d) be a bounded subspace of $(\mathbb{R}^n, d)$. Then there is $M > 0$ such that $A \subseteq [-M, M]^n$. Let $r \in \mathbb{R}^+$. Then $C = \left\{\frac{r}{2}k \;\middle|\; k \in \mathbb{Z}\right\}^n \cap [-M, M]^n$ is an r-net. □

12. Prove that every compact metrizable space is separable. Is every compact topological space separable?

Proof: Let $(S,\mathcal{T})$ be a compact metrizable space and suppose that $\mathcal{T}$ is induced by the metric d. For each $n \in \mathbb{N}$, let $\mathcal{C}_n = \left\{B_{\frac{1}{n}}(x) \;\middle|\; x \in S\right\}$. $\mathcal{C}_n$ is an open covering of S, and so, by compactness, there is a finite subcollection $\mathcal{D}_n \subseteq \mathcal{C}_n$ that still covers S. Let $D = \bigcup \mathcal{D}_n$. D is a countable union of countable sets, and so, D is countable. To see that D is dense, let U be a nonempty open set in S. Then there is an open ball $B_r(x) \subseteq U$. By the Archimedean Property of $\mathbb{R}$, there is $n \in \mathbb{N}$ such that $n > \frac{1}{r}$, or equivalently, $\frac{1}{n} < r$. Since $\mathcal{D}_n$ is an open covering of S, there is $x \in B_{\frac{1}{n}}(y)$ for some $y \in S$. Therefore, we have $x \in B_r(x) \cap B_{\frac{1}{n}}(y) \subseteq U \cap D$. This proves that D is a countable dense subset of S. □

A compact topological space does **not** need to be separable. For example, if we let $S = \mathbb{R}$ and we let $\mathcal{T} = \{\mathbb{R}\} \cup \{A \subseteq \mathbb{R} \mid 0 \notin A\}$. It is easily checked that $(\mathbb{R},\mathcal{T})$ is a topological space. To see that $(\mathbb{R},\mathcal{T})$ is compact, just observe that any open covering of $\mathbb{R}$ must contain $\mathbb{R}$ because $\mathbb{R}$ is the only open set containing 0. We now show that $(\mathbb{R},\mathcal{T})$ is **not** separable. To see this, let $D \subseteq \mathbb{R}$ be dense. If $x \neq 0$, then $\{x\}$ is open, and so, $\{x\} \cap D \neq \emptyset$. Therefore, $\mathbb{R} \setminus \{0\} \subseteq D$, showing that D is uncountable. Therefore, there is no countable dense subset of $\mathbb{R}$, and so, $(\mathbb{R},\mathcal{T})$ is not separable.

13. Prove that the one-point compactification of a locally compact T_2-space is a compact T_2-space.

Proof: Let's first verify that $\overline{\mathcal{T}}$ defines a topology on $\overline{S}$.

Since $\mathcal{T} \subseteq \overline{\mathcal{T}}$ and $\emptyset \in \mathcal{T}$, we have $\emptyset \in \overline{\mathcal{T}}$. Since $\emptyset$ is compact, $\overline{S} = S \cup \{\infty\} = (S \setminus \emptyset) \cup \{\infty\} \in \overline{\mathcal{T}}$.

Let $\boldsymbol{X} \subseteq \overline{\mathcal{T}}$. If $\boldsymbol{X} \subseteq \mathcal{T}$, then since $\mathcal{T}$ is a topology on S, $\bigcup \boldsymbol{X} \in \mathcal{T} \subseteq \overline{\mathcal{T}}$. Otherwise, let $\boldsymbol{W} = \{A \mid A \in \boldsymbol{X} \cap \mathcal{T}\}$ and let $\boldsymbol{Z} = \{S \setminus K \mid (S \setminus K) \cup \{\infty\} \in \boldsymbol{X}\}$.

Let $\boldsymbol{C} = \bigcap\{S \setminus A \mid A \in \boldsymbol{X} \cap \mathcal{T}\}$, and $\boldsymbol{L} = \bigcap\{K \mid (S \setminus K) \cup \{\infty\} \in \boldsymbol{X}\}$. $\boldsymbol{C}$ is an intersection of closed sets in S. Since $\mathcal{T}$ is a topology on S, $\boldsymbol{C}$ is closed in S. $\boldsymbol{L}$ is an intersection of compact sets. Since $\boldsymbol{L} \subseteq S$ and $(S, \mathcal{T})$ is a T_2-space, by Theorem 12.12, $\boldsymbol{L}$ is closed in S. Therefore, $\boldsymbol{C} \cap \boldsymbol{L}$ is closed in S. Since $\boldsymbol{C} \cap \boldsymbol{L}$ is a subset of a compact set (any K being used in the intersection $\boldsymbol{L}$ will work here), by Theorem 12.13, $\boldsymbol{C} \cap \boldsymbol{L}$ is compact. Now, $\bigcup \boldsymbol{X} = [\bigcup \boldsymbol{W} \cup \bigcup \boldsymbol{Z}] \cup \{\infty\} = [(S \setminus \boldsymbol{C}) \cup (S \setminus \boldsymbol{L})] \cup \{\infty\} = [S \setminus (\boldsymbol{C} \cap \boldsymbol{L})] \cup \{\infty\}$. Since $\boldsymbol{C} \cap \boldsymbol{L}$ is compact, $\bigcup \boldsymbol{X} \in \overline{\mathcal{T}}$. Finally, let $\boldsymbol{Y} \subseteq \overline{\mathcal{T}}$ with $\boldsymbol{Y}$ finite. If $\infty \notin U$ for some $U \in \boldsymbol{Y}$, then $\bigcap \boldsymbol{Y}$ is a finite intersection of sets in $\mathcal{T}$, and since $\mathcal{T}$ is a topology on S, $\bigcap \boldsymbol{Y} \in \mathcal{T} \subseteq \overline{\mathcal{T}}$. So, assume that every set in $\boldsymbol{Y}$ has the form $(S \setminus K) \cup \{\infty\}$ for some compact $K \subseteq S$. Let's say the compact sets are $K_0, \dots, K_n$. Then we have $\bigcap \boldsymbol{Y} = [(S \setminus K_0) \cap \cdots \cap (S \setminus K_n)] \cup \{\infty\} = [S \setminus (K_0 \cup \cdots \cup K_n)] \cup \{\infty\}$. Since a finite union of compact sets is compact, we see that $\bigcap \boldsymbol{Y} \in \overline{\mathcal{T}}$.

Let's now show that $(\overline{S}, \overline{\mathcal{T}})$ is a T_2-space. Let $x, y \in \overline{S}$. Since $(S, \mathcal{T})$ is a T_2-space, we can assume that one of these points is ∞, let's say $y = \infty$. Since S is locally compact, there is a compact set $K \subseteq S$ and an open set $U \subseteq K$ with $x \in U$. Then $(S \setminus K) \cup \{\infty\}$ is an open set containing ∞ and disjoint from U. Therefore, $(\overline{S}, \overline{\mathcal{T}})$ is a T_2-space.

Finally, we show that $(\overline{S}, \overline{\mathcal{T}})$ is a compact. To see this, let $\mathcal{C}$ be an open covering of $\overline{S}$. If $\overline{S} \in \mathcal{C}$, then $\{\overline{S}\}$ is a fnite subcover. So, we can assume that $\overline{S} \notin \mathcal{C}$. Since $\mathcal{C}$ is a covering of $\overline{S}$, there is $V \in \mathcal{C}$ with $\infty \in V$. Therefore, $\overline{S} \setminus V$ is a compact subspace of S. Now, $\overline{S} \setminus V$ is covered by the sets in $\mathcal{C}$, and so, there is a finite subset $\mathcal{D} \subseteq \mathcal{C}$ such that $\mathcal{D}$ till covers $\overline{S} \setminus V$. Then $\mathcal{D} \cup \{V\}$ is the desired finite subcover. □

LEVEL 5

14. Prove that a locally compact T_2-space is a T_3-space.

Proof: Let $(S, \mathcal{T})$ be a locally compact T_2-space, let $x \in S$, and let A be a closed set $A \subseteq S \setminus \{x\}$. Since $(S, \mathcal{T})$ is locally compact, there is a compact subspace $(K, \mathcal{T}_K)$ such that K contains a neighborhood W of x. By Theorem 12.12, K is closed in S. So, $A \cap K$ is closed in S. By Theorem 12.13, $(A \cap K, \mathcal{T}_{A \cap K})$ is a compact subspace of $(S, \mathcal{T})$. By Problem 10 above, there are open sets V_1 and V_2 with $x \in V_1$ and $A \cap K \subseteq V_2$. Let $U = V_1 \cap W$ and $V = V_2 \cup (S \setminus K)$. Then U and V are open and disjoint, $x \in U$ and $A \subseteq V_0$. So, $(S, \mathcal{T})$ is a T_3-space. □

15. Prove Lebesgue's Covering Lemma (Theorem 12.17).

Proof: Let (S, d) be a sequentially compact metric space and let $\mathcal{C}$ be an open cover of S. If every subset of S is contained in a member of $\mathcal{C}$, then every positive real number is a Lebesgue number for $\mathcal{C}$. So, we may assume that there exist subsets of S that are not contained in any member of $\mathcal{C}$. Let $c = \inf\{\text{diam } A \mid A \subseteq S \wedge A \text{ is not contained in any member of } \mathcal{C}\}$. If $x = +\infty$, then once again every positive real number is a Lebesgue number for $\mathcal{C}$. So, we may assume that c is a nonnegative real number. Also, if $c > 0$, then c is a Lebesgue number for $\mathcal{C}$. Assume toward contradiction that $c = 0$. Then for each $in\ \mathbb{Z}^+$, there is a subset A_n of S such that A_n is not contained in any member of $\mathcal{C}$ and $0 < \text{diam } A_n < \frac{1}{n+1}$. Let $s_n \in A_n$ for each $n \in \mathbb{N}$. Since (S, d) is sequentially compact, the sequence (s_n) has a subsequence that converges to some $s \in S$. Since $\mathcal{C}$ is an open cover of S, there is $U \in \mathcal{C}$ such that $s \in U$. Since U is open, there is $r \in \mathbb{R}^+$ such that $B_r(s) \subseteq U$. Since $s_n \to s$, there are infinitely many natural number n for which $s_n \in B_{\frac{r}{2}}(s)$. Choose $K \in \mathbb{N}$ such that $s_K \in B_{\frac{r}{2}}(s)$ and $\frac{1}{K} < \frac{r}{2}$. Since $\text{diam } A_K < \frac{1}{K} < \frac{r}{2}$, we have $A_K \subseteq B_r(s)$ (check this!), and therefore, $A_K \subseteq U$, contradicting our assumption that A_K is not contained in any member of $\mathcal{C}$. □

16. Let $(S, \mathcal{T})$ be a topological space. Prove each of the following:

(i) A is nowhere dense in S if and only if for any open set $U \subseteq S$, there is a nonempty open set V with $V \subseteq U$ and $V \cap A = \emptyset$.

(ii) The union of finitely many sets that are nowhere dense in S is nowhere dense in S.

(iii) A is nowhere dense in S if and only if $(S \setminus A)^\circ$ is dense in S.

Proofs:

(i) First suppose that A is nowhere dense in S and let $U \subseteq S$ be a nonempty open set. If $U \subseteq \overline{A}$, then since U is open, $U \subseteq (\overline{A})^\circ$, contradicting that $(\overline{A})^\circ = \emptyset$. So, $V = U \cap (S \setminus \overline{A}) \neq \emptyset$. Therefore, V is a nonempty open set with $V \subseteq U$. Since $A \subseteq \overline{A}$, we have $S \setminus \overline{A} \subseteq S \setminus A$. Since $V \subseteq S \setminus \overline{A}$, we have $V \subseteq S \setminus A$, and therefore, $V \cap A = \emptyset$.

Conversely, assume that A is **not** nowhere dense. Then there is $U \subseteq \overline{A}$ with $U \neq \emptyset$ and U open in S. Let $V \subseteq U$ be a nonempty open set. Since $V \subseteq \overline{A}$, $V \cap A \neq \emptyset$. So, U is an open set that does **not** contain an open set that does not intersect A. □

(ii) Let's first show that the union of two nowhere dense sets is nowhere dense. So, let A and B be nowhere dense in S. Let $U \subseteq S$ be a nonempty open set. Since A is nowhere dense, by (i), there is a nonempty open set V such that $V \subseteq U$ and $V \cap A = \emptyset$. Similarly, since B is nowhere dense, by (i), there is a nonempty open set W such that $W \subseteq V$ and $W \cap B = \emptyset$. It follows that $W \subseteq U$ and $W \cap (A \cup B) = (W \cap A) \cup (W \cap B) \subseteq (V \cap A) \cup (W \cap B) = \emptyset \cup \emptyset = \emptyset$. By (i) again, $A \cup B$ is nowhere dense.

We proceed by induction. Assuming that the union of k sets that are nowhere dense in S is nowhere dense in S, let $A_0, \dots, A_k$ be sets that are nowhere dense in S. By the inductive assumption, $A_0 \cup \cdots \cup A_{k-1}$ is nowhere dense in S. Finally, by the base case, we have

$$A_0 \cup \cdots \cup A_k = (A_0 \cup \cdots \cup A_{k-1}) \cup A_k,$$

and so, $A_0 \cup \cdots \cup A_k$ is nowhere dense in S. □

(iii) Let $(S, \mathcal{T})$ be a topological space. First note that for any subset $X \subseteq S$, we have the following equalities: $S \setminus \overline{X} = (S \setminus X)^\circ$ and $S \setminus X^\circ = \overline{S \setminus X}$.

To see that $S \setminus \overline{X} = (S \setminus X)^\circ$, note that we have

$$x \in S \setminus \overline{X} \Leftrightarrow x \notin \overline{X} \Leftrightarrow \text{There is an open set } U \text{ with } x \in U \text{ and } U \cap X = \emptyset$$
$$\Leftrightarrow \text{There is an open set } U \text{ with } x \in U \text{ and } U \subseteq S \setminus X \Leftrightarrow x \in (S \setminus X)^\circ.$$

To see that $S \setminus X^\circ = \overline{S \setminus X}$, note that we have

$$x \in S \setminus X^\circ \Leftrightarrow x \notin X^\circ \Leftrightarrow \text{For every open set } U \text{ with } x \in U, U \cap (S \setminus X) \neq \emptyset \Leftrightarrow x \in \overline{S \setminus X}.$$

Now, let $A \subseteq S$. We have

$$A \text{ is nowhere dense in } S \Leftrightarrow (\overline{A})^\circ = \emptyset \Leftrightarrow S \setminus (\overline{A})^\circ = S \Leftrightarrow \overline{S \setminus \overline{A}} = S \Leftrightarrow \overline{(S \setminus X)^\circ} = S.$$

This proves the result. □

Problem Set 13

LEVEL 1

1. Let $(A, \mathcal{T})$ and $(B, \mathcal{U})$ be topological spaces and let $f: A \to B$. Prove that f is continuous if and only if for each C that is closed in B, the set $f^{-1}[C]$ is closed in A.

Proof: Let $(A, \mathcal{T})$ and $(B, \mathcal{U})$ be topological spaces and let $f: A \to B$. First assume that f is continuous. Let C be closed in B. Then $B \setminus C$ is open in B. Since f is continuous, $f^{-1}[B \setminus C]$ is open in A. We will show that $f^{-1}[B \setminus C] = A \setminus f^{-1}[C]$. To see this, note that $x \in f^{-1}[B \setminus C]$ if and only if $f(x) \in B \setminus C$ if and only if $f(x) \in B$ and $f(x) \notin C$ if and only if $x \in f^{-1}[B] = A$ and $x \notin f^{-1}[C]$ if and only if $x \in A \setminus f^{-1}[C]$. It follows that $A \setminus f^{-1}[C]$ is open in A, and therefore, $f^{-1}[C]$ is closed in A.

Conversely, assume that for each C that is closed in B, the set $f^{-1}[C]$ is closed in A. Let U be open in B. Then $B \setminus U$ is closed in B. By our assumption, $f^{-1}[B \setminus U]$ is closed in A. By the same argument given in the first paragraph above, we have $f^{-1}[B \setminus U] = A \setminus f^{-1}[U]$. So, $A \setminus f^{-1}[U]$ is closed in A. Therefore, $f^{-1}[U]$ is open in A. Since U was an arbitrary open set in B, f is continuous. □

2. Prove that the one-point compactification of a locally compact T_2-space is a Tychonoff space.

Proof: Let $(\overline{S}, \overline{\mathcal{T}})$ be the one-point compactification of $(S, \mathcal{T})$. By Problem 13 in Problem Set 12, $(\overline{S}, \overline{\mathcal{T}})$ is a compact T_2-space. By Problem 10 in Problem Set 12, $(\overline{S}, \overline{\mathcal{T}})$ is a T_4-space. By Corollary 13.10, $(\overline{S}, \overline{\mathcal{T}})$is a Tychonoff space. □

LEVEL 2

3. Prove that $\mathbb{C}^n$ and $\mathbb{R}^{2n}$ with their standard topologies are topologically equivalent.

Proof: Define $f: \mathbb{C}^n \to \mathbb{R}^{2n}$ by $f(a_1 + b_1 i, a_2 + b_2 i, \dots, a_n + b_n i) = (a_1, b_1, a_2, b_2, \dots, a_n, b_n)$. If

$$f(a_1 + b_1 i, a_2 + b_2 i, \dots, a_n + b_n i) = f(c_1 + d_1 i, c_2 + d_2 i, \dots, c_n + d_n i),$$

then we have $(a_1, b_1, a_2, b_2, \dots, a_n, b_n) = (c_1, d_1, c_2, d_2, \dots, c_n, d_n)$. So, $a_1 = c_1$, $b_1 = d_1$,...,$b_n = d_n$. Therefore, $(a_1 + b_1 i, a_2 + b_2 i, \dots, a_n + b_n i) = (c_1 + d_1 i, c_2 + d_2 i, \dots c_n + d_n i)$. It follows that f is injective.

If $(x_1, x_2, \dots, x_{2n}) \in \mathbb{R}^{2n}$, then $f(x_1 + x_2 i, x_3 + x_4 i, \dots, x_{2n-1} + x_{2n} i) = (x_1, x_2, \dots, x_{2n})$. This shows that f is surjective.

f is continuous because $f^{-1}[B_r(x_1, x_2, \dots, x_{2n})] = B_r(x_1 + x_2 i, x_3 + x_4 i, \dots, x_{2n-1} + x_{2n} i)$ for any open ball $B_r(x_1, x_2, \dots, x_{2n})$ in $\mathbb{R}^{2n}$.

Similarly, f^{-1} is continuous because $f[B_r(x_1 + y_1 i, x_2 + y_2 i, \dots, x_n + y_n i)] = B_r(x_1, y_1, \dots, x_n, y_n)$. □

4. Let $(A,\mathcal{T})$ and $(B,\mathcal{U})$ be topological spaces, let $x \in A$, and let $f: A \to B$. Prove that if f is continuous at x, then then for every sequence x_n converging to x, the sequence $f(x_n)$ converges to $f(x)$. Then prove that if $(A,\mathcal{T})$ is a first-countable space, then the converse holds.

Proof: Let $(A,\mathcal{T})$ and $(B,\mathcal{U})$ be topological spaces, let $x \in A$, and let $f: A \to B$. First assume that f is continuous at x. Let (x_n) be a sequence in A such that $x_n \to x$. Let V be an open set containing $f(x)$. Since f is continuous at x there is an open set U containing x such that $f[U] \subseteq V$. Since $x_n \to x$, by Theorem 11.4, there is $K \in \mathbb{N}$ such that $n > K$ implies $x_n \in U$. So, $n > K$ implies $f(x_n) \in f[U]$. Since $f[U] \subseteq V$, it follows that $n > K$ implies $f(x_n) \in V$. Therefore, $f(x_n) \to f(x)$.

Conversely, suppose that $(A,\mathcal{T})$ is first countable and f is **not** continuous at x. Then there is an open set V with $f(x) \in V$ such that for any open set U containing x, $f[U] \nsubseteq V$. Let $\mathcal{B} = \{U_n \mid n \in \mathbb{N}\}$ be a countable basis at x. For each $n \in \mathbb{N}$, let $W_n = \bigcap\{U_k \mid k \leq n\}$. Let $n \in \mathbb{N}$. Since $x \in U_k$ for each $k = 0, 1, \dots, n$, $x \in W_n$. Since W_n is a finite intersection of open sets, it is open. For each $n \in \mathbb{N}$, let $x_n \in W_n$ with $f(x_n) \notin V$ (we can find x_n because $f[W_n] \nsubseteq V$). As in the proof of Theorem 10.14, $x_n \to x$. Since $f(x_n) \notin V$ for all $n \in \mathbb{N}$, $f(x_n) \nrightarrow f(x)$. □

5. Let $(A,\mathcal{T})$ and $(B,\mathcal{U})$ be metrizable topological spaces. Let $f, g: A \to B$ be continuous and let $C \subseteq A$. Suppose that $f(x) = g(x)$ for all $x \in A$. Prove that $f(x) = g(x)$ for all $x \in \overline{A}$.

Proof: Let $x \in \overline{A}$. Then there is a sequence (x_n) in A such that $x_n \to x$ (simply choose $x_n \in B_{\frac{1}{n}}(x)$ for each $n \in \mathbb{N}$). Since f is continuous, $f(x_n) \to f(x)$. Since g is continuous, $f(x_n) = g(x_n) \to g(x)$. So, $f(x_n)$ converges to both $f(x)$ and $g(x)$. Since metrizable spaces are T_2-spaces, by Theorem 10.9, $f(x) = g(x)$. □

LEVEL 3

6. Let $(K,\mathcal{T})$ and $(L,\mathcal{U})$ be topological spaces with $(K,\mathcal{T})$ compact and let $f: K \to L$ be a homeomorphism. Prove that $(L,\mathcal{U})$ is compact.

Proof: Let $\mathcal{C}$ be an open covering of L. Since f is continuous and bijective, $\mathcal{D} = \{f^{-1}[B] \mid B \in \mathcal{C}\}$ is an open covering of K. Since $(K,\mathcal{T})$ is compact, there is a finite subcollection $\mathcal{E} \subseteq \mathcal{D}$ that covers K. Since f^{-1} is continuous and bijective, $\mathcal{H} = \{f[f^{-1}[B]] \mid f^{-1}[B] \in \mathcal{E}\}$ covers L. By part Problem 5 from Problem Set 4, $f[f^{-1}[B]] \subseteq B$. Since $\mathcal{H}$ covers L, so does $\mathcal{J} = \{B \mid f^{-1}[B] \in \mathcal{E}\}$. Finally, $\mathcal{J} \subseteq \mathcal{C}$ because if $B \in \mathcal{J}$, then $f^{-1}[B] \in \mathcal{E}$. So, $f^{-1}[B] \in \mathcal{D}$. Therefore, $B \in \mathcal{C}$. □

Note: Since f is surjective, we actually have $f[f^{-1}[B]] = B$. To see this, first note that by Problem 1 above, we have $f[f^{-1}[B]] \subseteq B$. For the other inclusion, let $y \in B$. Since f is surjective, there is $x \in f^{-1}[B]$ with $f(x) = y$. Then $y = f(x) \in f[f^{-1}[B]]$. Since $y \in B$ was arbitrary, we have $B \subseteq f[f^{-1}[B]]$. Since $f[f^{-1}[B]] \subseteq B$ and $B \subseteq f[f^{-1}[B]]$, we have $f[f^{-1}[B]] = B$.

7. Prove that the image of a Cauchy sequence under a uniformly continuous function is a Cauchy sequence. If we replace "uniformly continuous" by "continuous," is the result still true?

Proof: Let (A, d) and (B, ρ) be metric spaces, let $f: A \to B$ be uniformly continuous, let (x_n) be a Cauchy sequence in A, and let $\epsilon > 0$. Since f is uniformly continuous, there is $\delta > 0$ such that for all $x, y \in A$, $d(x, y) < \delta \to \rho(f(x), f(y)) < \epsilon$. Since (x_n) is a Cauchy sequence, there is $K \in \mathbb{N}$ such that $m \geq n > K$ implies $d(x_m, x_n) < \delta$. Now, assume that $m \geq n > K$. Then we have $d(x_m, x_n) < \delta$. Therefore, $\rho(f(x_m), f(x_n)) < \epsilon$. It follows that $(f(x_n))$ is a Cauchy sequence in B. □

If we replace "uniformly continuous" by "continuous," the result is false. Define $f: (0, 1) \to \mathbb{R}$ by $f(x) = \frac{1}{x}$. Then $\left(\frac{1}{n}\right)$ is a Cauchy sequence in $(0, 1)$, but $f\left(\frac{1}{n}\right) = n$, and (n) is not a Cauchy sequence in $\mathbb{R}$.

8. Let $(S, \mathcal{T})$ be a T_4-space, let A and B be disjoint closed subsets of S, and let a and b be real numbers with $a < b$. Prove that there is a continuous function $f: S \to [a, b]$ such that $f[A] = \{a\}$ and $f[B] = \{b\}$.

Proof: Let $(S, \mathcal{T})$ be a T_4-space, let A and B be disjoint closed subsets of S, and let a and b be real numbers with $a < b$. By Urysohn's Lemma (Theorem 13.9), there is a continuous function $g: S \to [0, 1]$ such that $g[A] = \{0\}$ and $g[B] = \{1\}$. Define $f: S \to [a, b]$ by $f(x) = (b - a)g(x) + a$. Then we have $f[A] = (b - a)g[A] + a = (b - a)(0) + a = a$, $\quad f[B] = (b - a)g[B] + a = (b - a)(1) + a = b$, and f is continuous by a standard continuity argument. □

LEVEL 4

9. Let $(A, \mathcal{T})$ and $(B, \mathcal{U})$ be topological spaces and let $f: A \to B$. Prove that f is continuous if and only if for each $X \subseteq A$, $f[\overline{X}] \subseteq \overline{f[X]}$.

Proof: Let $(A, \mathcal{T})$ and $(B, \mathcal{U})$ be topological spaces and let $f: A \to B$. First assume that f is continuous. Let $f(x) \in f[\overline{X}]$ and let U be an open set containing $f(x)$. Then $f^{-1}[U]$ is an open set containing x. Since $x \in \overline{X}$, there is $z \in \overline{X} \cap f^{-1}[U]$. Since $z \in \overline{X}$, $f(z) \in f[\overline{X}]$. Since $z \in f^{-1}[U]$, $f(z) \in f[f^{-1}[U]]$. By Problem 5 from Problem Set 4, $f[f^{-1}[U]] \subseteq U$. So, $f(z) \in U$. Thus, $f(z) \in f[\overline{X}] \cap U$. By part 5 of Theorem 9.4, $f(x) \in f[\overline{X}]$. Since $f(x) \in f[\overline{X}]$ was arbitrary, $f[\overline{X}] \subseteq \overline{f[X]}$.

Conversely, assume that for each $X \subseteq A$, $f[\overline{X}] \subseteq \overline{f[X]}$. Let C be a closed set in B and let $D = f^{-1}[C]$. By Problem 5 from Problem Set 4, $f[D] = f[f^{-1}[C]] \subseteq C$. Now, if $x \in \overline{D}$, then we have $f(x) \in f[\overline{D}]$. By our assumption, $f[\overline{D}] \subseteq \overline{f[D]}$. So, $f(x) \in \overline{f[D]} \subseteq \overline{C} = C$ (because C is closed in B). Therefore, we have $x \in f^{-1}[C] = D$. Since $x \in \overline{D}$ was arbitrary, $\overline{D} \subseteq D$. Since $D \subseteq \overline{D}$ is always true (by part 1 of Theorem 9.4), we have $D = \overline{D}$. By part 4 of Theorem 9.4, $D = f^{-1}[C]$ is closed in A. By Problem 1 above, f is continuous. □

10. Prove that an injective continuous function from a compact space onto a T_2-space is a homeomorphism.

Proof: Let $(S_1, \mathcal{T}_1)$ be a compact space, let $(S_2, \mathcal{T}_2)$ be a T_2-space, and let $f: S_1 \to S_2$ be a bijective continuous function.

We first prove that if K is a compact subspace of S_1, then $f[K]$ is a compact subspace of S_2. So, let K be a compact subspace of S_1, and let $\mathcal{C}$ be an open covering of $f[K]$ by open sets in S_2. Let $\mathcal{D} = \{f^{-1}[U] \mid U \in \mathcal{C}\}$. Since f is continuous, every element of $\mathcal{D}$ is open in S_1. If $x \in K$, then $f(x) \in f[K]$. Therefore, there is $U \in \mathcal{C}$ with $f(x) \in U$. It follows that $x \in f^{-1}[U]$. Since $x \in K$ was arbitrary, we see that $\mathcal{D}$ is an open covering of K by open sets in S_1. By the compactness of K and Lemma 12.10, there is a finite subcollection $\mathcal{E}$ of $\mathcal{D}$ that still covers K. Let $\mathcal{F} = \{U \in \mathcal{C} \mid f^{-1}[U] \in \mathcal{E}\}$. Then U is a finite subcollection of $\mathcal{C}$. Let $f(x) \in f[K]$. Then $x \in K$. Since $\mathcal{E}$ covers K, there is $f^{-1}[U] \in \mathcal{E}$ such that $x \in f^{-1}[U]$. So, $f(x) \in U$ and $U \in \mathcal{F}$. It follows that $\mathcal{F}$ is a finite subcollection of $\mathcal{C}$ that still covers $f[K]$. So, $f[K]$ is a compact subspace of S_2.

Now, let C be a closed set in S_1. If $C = \emptyset$, then $f[C] = \emptyset$, which is closed. Assume that $C \neq \emptyset$. By Theorem 12.13, C is compact in S_1. By the preceding paragraph, $f[C]$ is compact in S_2. By Theorem 12.12, $f[C]$ is closed in S_2. By Problem 1 above, f^{-1} is continuous. □

LEVEL 5

11. Let $(K, \mathcal{T})$ and $(B, \mathcal{U})$ be metrizable topological spaces with $(K, \mathcal{T})$ compact and let $f: K \to B$ be continuous. Prove that f is uniformly continuous.

Proof: Let $(K, \mathcal{T})$ and $(B, \mathcal{U})$ be metrizable topological spaces with $(K, \mathcal{T})$ compact induced by metrics d and ρ, respectively. Let $f: K \to B$ be continuous, let $\epsilon > 0$, and let $\mathcal{C} = \left\{B_{\frac{\epsilon}{2}}(y; \rho) \,\middle|\, y \in B\right\}$. Then $\mathcal{C}$ is an open covering of B. The set $\mathcal{D} = \left\{f^{-1}\left[B_{\frac{\epsilon}{2}}(y; \rho)\right] \,\middle|\, y \in B\right\}$ is an open covering of K. Let δ be a Lebesgue number for $\mathcal{D}$. If $a, b \in K$ with $d(a, b) < \delta$, then diam $\{a, b\} < \delta$. So, there is $y \in B$ such that $\{a, b\} \subseteq f^{-1}\left[B_{\frac{\epsilon}{2}}(y; \rho)\right]$. Therefore, $\{f(a), f(b)\} \in f\left[f^{-1}\left[B_{\frac{\epsilon}{2}}(y; \rho)\right]\right] \subseteq B_{\frac{\epsilon}{2}}(y; \rho)$. It follows that $\rho(f(a), f(b)) \leq \rho(f(a), y) + \rho(y, f(b)) < \frac{\epsilon}{2} + \frac{\epsilon}{2} = \epsilon$. □

12. Let $(A, \mathcal{T})$ be a metrizable topological space and let $(B, \mathcal{U})$ be a completely metrizable topological space. Let $C \subseteq A$ be dense in A and suppose that $f: C \to B$ is uniformly continuous. Prove that f can be extended uniquely to a uniformly continuous function $g: A \to B$. If we replace "uniformly continuous" by "continuous," is the result still true?

Proof: Let $(A, \mathcal{T})$ be a metrizable topological space and let $(B, \mathcal{U})$ be a completely metrizable topological spaces induced by metrics d and ρ, respectively. Let $C \subseteq A$ be dense in A and let $f: C \to B$ be uniformly continuous. If $C = A$, we can let $g = f$. So, assume that $C \neq A$. If $x \in C$, let $g(x) = f(x)$. Now, let $x \in A \setminus C$. Since C is dense in A, $\overline{C} = A$. Therefore, there is a sequence (x_n) in C such that $x_n \to x$. By Theorem 11.5, (x_n) is a Cauchy sequence in C. By Problem 7 above, $(f(x_n))$ is a Cauchy sequence in B. Since (B, ρ) is a complete metric space, $(f(x_n))$ converges to some point y in B. Let $g(x) = y$. If (z_n) is another Cauchy sequence in C such that $z_n \to x$, then $d(x_n, z_n) \to 0$. Since f is uniformly continuous, $\rho(f(x_n), f(z_n)) \to 0$. Therefore, $f(z_n) \to y = g(x)$.

We need to show that g is uniformly continuous. To this end, let $\epsilon > 0$. Since f is uniformly continuous, there is $\delta > 0$ such that for all $x, z \in C$, $d(x,z) < \delta \to \rho(f(x), f(z)) < \epsilon$. Suppose that $a, b \in A$ with $d(a,b) < \delta$. Let $(a_n), (b_n)$ be sequences in C such that $a_n \to a$ and $b_n \to b$. We then have that $d\{a_n, b_n) \leq d(a_n, a) + d(a,b) + d(b, b_n)$. Since $d(a_n, a) \to 0$, $d(b, b_n) \to 0$, and $d(a,b) < \delta$, we can find $K \in \mathbb{N}$ such that for all $n > K$, $d(a_n, b_n) < \delta$. So, $n > K \to \rho(f(a_n), f(b_n)) < \epsilon$. From this it is easy to show that $\rho(g(a), g(b)) \leq \epsilon$.

The proof that g is unique is straightforward.

If we replace "uniformly continuous" by "continuous," the result is still true. The proof is similar. □

Problem Set 14

LEVEL 1

1. Prove that a topological space $(S, \mathcal{T})$ is disconnected if and only if there is a surjective continuous function $f: S \to \{0, 1\}$, where $\{0, 1\}$ is given the discrete topology.

Proof: First assume that $(S, \mathcal{T})$ is disconnected and let (U, V) be a disconnection of S. Define $f: S \to \{0, 1\}$ by $f(x) = \begin{cases} 0 & \text{if } x \in U. \\ 1 & \text{if } x \in V. \end{cases}$ Since U and V are nonempty, f is surjective. $f^{-1}[\{0\}] = U$, which is open in S and $f^{-1}[\{1\}] = V$, which is also open in S. So, f is continuous.

Conversely, suppose that $f: S \to \{0, 1\}$ is surjective and continuous. Let $U = f^{-1}[\{0\}]$ and $V = f^{-1}[\{1\}]$. Since f is continuous and $\{0\}$ and $\{1\}$ are open in the discrete topology of $\{0, 1\}$, U and V are both open in S. Since f is surjective, U and V are both nonempty. If $x \in U \cap V$, then $f(x) = 0$ and $f(x) = 1$, which is a contradiction. So, $U \cap V = \emptyset$. Finally, if $x \in S$, then $f(x) \in \{0, 1\}$. If $f(x) = 0$, then $x \in U$. If $f(x) = 1$, then $x \in V$. So, $x \in U$ or $x \in V$. Therefore, $x \in U \cup V$. Since $x \in S$ was arbitrary, $S = U \cup V$. It follows that (U, V) is a disconnection of S. So, $(S, \mathcal{T})$ is disconnected. □

2. Let $(S, \mathcal{T})$ be a topological space and define the relation $\sim$ on S by $x \sim y$ if and only if $x, y \in A$ for some connected subspace $(A, \mathcal{T}_A)$ of $(S, \mathcal{T})$. Prove that $\sim$ is an equivalence relation on S.

Proof: Let $x \in S$. Then $(\{x\}, \mathcal{T}_{\{x\}})$ is a connected subspace of $(S, \mathcal{T})$ with $x \in A$. So, $x \sim x$, and therefore, $\sim$ is reflexive.

Let $x, y \in S$ with $x \sim y$. Then $x, y \in A$ for some connected subspace $(A, \mathcal{T}_A)$ of $(S, \mathcal{T})$. So, $y, x \in A$ for some connected subspace $(A, \mathcal{T}_A)$ of $(S, \mathcal{T})$. Thus, $y \sim x$, and so, $\sim$ is symmetric.

Let $x, y, z \in S$ with $x \sim y$ and $y \sim z$. Then there are connected subspaces $(A, \mathcal{T}_A)$ and $(B, \mathcal{T}_B)$ of $(S, \mathcal{T})$ such that $x, y \in A$ and $y, z \in B$. Since $y \in A \cap B$, by Theorem 14.7, $(A \cup B, \mathcal{T}_{A \cup B})$ is a connected subspace of $(S, \mathcal{T})$ with $x, y, z \in A \cup B$. So, $x \sim z$, and therefore, $\sim$ is transitive.

Since $\sim$ is reflexive, symmetric, and transitive, $\sim$ is an equivalence relation. □

LEVEL 2

3. Let $(S_1, \mathcal{T}_1)$ and $(S_2, \mathcal{T}_2)$ be connected topological spaces. Prove that $(S_1 \times S_2, \mathcal{T})$ is connected, where $\mathcal{T}$ is the product topology on $S_1 \times S_2$.

Proof: Let $(S_1, \mathcal{T}_1)$ and $(S_2, \mathcal{T}_2)$ be connected topological spaces and let $f: S_1 \times S_2 \to \{0, 1\}$ be continuous (here $\{0, 1\}$ is given the discrete topology). By Problem 1 above, it suffices to show that f is constant. Let $a \in S_1$ and define $g: S_2 \to \{0, 1\}$ by $g(y) = f(a, y)$. Then g is continuous. Since $(S_2, \mathcal{T}_2)$ is connected, by Problem 1 again, g is constant. Similarly, for each $b \in S_2$, the function $h: S_1 \to \{0, 1\}$ defined by $h(x) = f(x, b)$ is constant. Let $(a, b), (c, d) \in S_1 \times S_2$. Then $f(a, b) = f(a, d) = f(c, d)$. So, f is constant. □

4. Let $(S,\mathcal{T})$ be a topological space. Prove that $(S,\mathcal{T})$ is locally connected if and only if each point of S has a basis $\mathcal{B}$ consisting of connected subspaces of S ($\mathcal{B}$ is a basis at x if (i) $V \in \mathcal{B} \to x \in V$; and (ii) for any open set U containing x, there is a $V \in \mathcal{B}$ such that $V \subseteq U$).

Proof: First assume that $(S,\mathcal{T})$ is locally connected, let $\mathcal{G}$ be a basis for $\mathcal{T}$, and let $x \in S$. For each $U \in \mathcal{T}$ with $x \in U$, let V_U be an open set containing x such that $(V,\mathcal{T}_V)$ is connected and $V_U \subseteq U$. Let $\mathcal{B} = \{V_U \mid U \in \mathcal{T} \wedge x \in U\}$. Then clearly $\mathcal{B}$ is a basis of x consisting of connected subspaces of S.

Conversely, suppose that each point of S has a basis consisting of connected subspaces of S. Let $x \in S$ and let U be an open set containing x. Let $\mathcal{B}$ be a basis of x consisting of connected subspaces of S. Then by (ii), there is $V \in \mathcal{B}$ such that $V \subseteq U$. □

5. Prove that local connectedness is a topological property.

Proof: Let $(A,\mathcal{T})$ and $(B,\mathcal{U})$ be topological spaces with $(A,\mathcal{T})$ locally connected, let $f: A \to B$ be a homeomorphism, let $y \in B$ and let W be an open set containing y. Then $y = f(x)$ for some $x \in A$ and $U = f^{-1}[W]$ is an open set containing x. Since $(A,\mathcal{T})$ is locally connected, there is an open set V containing x such that $(V,\mathcal{T}_V)$ is connected and $V \subseteq U$. Since f^{-1} is continuous, $f[V]$ is open in B. Since $x \in V$, $f(x) \in f[V]$. Also, $f[V] \subseteq f[U] = f\left[f^{-1}[W]\right] = W$. Finally, since $(V,\mathcal{T}_V)$ is connected and connectedness is a topological property, $\left(f[V],\mathcal{T}_{f[V]}\right)$ is connected. □

LEVEL 3

6. Prove that the components of a totally disconnected space are its points.

Proof: Let $(S,\mathcal{T})$ be a totally disconnected topological space and let $(A,\mathcal{T}_A)$ be a subspace of $(S,\mathcal{T})$ such that $|A| \geq 2$. Let $x, y \in A$ with $x \neq y$. Since $(S,\mathcal{T})$ is totally disconnected, there is a disconnection (U,V) of S with $x \in U$ and $y \in V$. Then $A = (A \cap U, A \cap V)$ is a disconnection of A with $x \in A \cap U$ and $y \in A \cap V$. □

7. Prove that an arbitrary product of totally disconnected spaces is totally disconnected.

Proof: Let K be an index set, let $(S_k,\mathcal{T}_k)$ be totally disconnected for each $k \in K$, and assume that $X \subseteq \prod S_k$ is connected. Then for each $k \in K$, $X_k = \{x_k \in S_k \mid \mathbf{x} \in X\}$ is connected. Since S_k is totally disconnected, $X_k = \{a_k\}$ for some $a_k \in S_k$. So, $X = \{\mathbf{a}\}$. So, $\prod S_k$ is totally disconnected. □

8. Let $(S,\mathcal{T})$ be a T_2-space and let $\mathcal{B}$ be a basis for $\mathcal{T}$ such that each set in $\mathcal{B}$ is both open and closed. Prove that $(S,\mathcal{T})$ is totally disconnected.

Proof: Let $(S,\mathcal{T})$ be a T_2-space and let $\mathcal{B}$ be a basis for $\mathcal{T}$ such that each set in $\mathcal{B}$ is both open and closed and let $x, y \in S$ with $x \neq y$. Since $(S,\mathcal{T})$ be a T_2-space, there are open sets U, V in S such that $x \in U$, $y \in V$, and $U \cap V = \emptyset$. Let $W \in \mathcal{B}$ with $x \in W$ and $W \subseteq U$. Then $(W, S \setminus W)$ is a disconnection of S with $x \in W$ and $y \in S \setminus W$. Since $x, y \in S$ were arbitrary, $(S,\mathcal{T})$ is totally disconnected. □

LEVEL 4

9. Consider the Cantor set C as a subspace of $\mathbb{R}$ with the standard topology. Prove that $(C, \mathcal{T}_C)$ is totally disconnected.

Proof: Let $x, y \in C$ with $x \neq y$. Let $K \in \mathbb{N}$ be such that $|x - y| > \frac{1}{3^K}$. Then x and y belong to different intervals in C_K. There is at least one interval between x and y that is disjoint from C_K, and therefore, from C. Let z be an element of that interval. Then $\big((-\infty, z) \cap C, (z, \infty) \cap C\big)$ is a disconnection of C separating x and y. □

10. Let $(S, \mathcal{T})$ be a topological space, let $(A, \mathcal{T}_A)$ be a connected subspace of $(S, \mathcal{T})$, and let $(B, \mathcal{T}_B)$ be a subspace of $(S, \mathcal{T})$ such that $A \subseteq B \subseteq \overline{A}$. Prove that $(B, \mathcal{T}_B)$ is connected.

Proof: Let $(S, \mathcal{T})$ be a topological space, let $(A, \mathcal{T}_A)$ be a connected subspace of $(S, \mathcal{T})$, and let $(B, \mathcal{T}_B)$ be a subspace of $(S, \mathcal{T})$ such that $A \subseteq B \subseteq \overline{A}$. Assume toward contradiction that B is disconnected. Let $(U \cap B, V \cap B)$ be a disconnection of B. Since A is connected and $A \subseteq B = (U \cap B) \cup (V \cap B)$, we have $A \subseteq U \cap B$ or $A \subseteq V \cap B$. Without loss of generality, assume that $A \subseteq U \cap B$. Therefore, $A \cap V = A \cap (V \cap B) = \emptyset$. So, $\overline{A} \cap V = \emptyset$. Since $B \subseteq \overline{A}$, $V \cap B = \emptyset$, contradcting that $(U \cap B, V \cap B)$ is a disconnection of B. □

11. Let $(S, \mathcal{T})$ be a topological space. Prove each of the following:

    (i) Every $x \in S$ is contained in exactly one component of S.

    (ii) If $(A, \mathcal{T}_A)$ is a connected subspace of $(S, \mathcal{T})$, then A is contained in a component of S.

    (iii) If $(A, \mathcal{T}_A)$ is a connected subspace of $(S, \mathcal{T})$ and A is both open and closed in S, then A is a component of S.

    (iv) Every component of S is closed in S.

Proofs:

(i) Let $x \in S$. Since the equivalence classes of $\sim$ form a partition of S, x is in exactly one component of S. □

(ii) Let $(A, \mathcal{T}_A)$ be a connected subspace of $(S, \mathcal{T})$. Let $B = \{x \mid \exists y \in A(x \sim y)\}$. If $x \in A$, then since $x \sim x$ (because $\sim$ is reflexive), $x \in B$. Therefore, $A \subseteq B$. Suppose toward contradiction that $(B, \mathcal{T}_B)$ is disconnected and let (U, V) be a disconnection of B. Let $a \in U$ and $b \in V$. Since $a, b \in B$, tehre are $y, z \in A$ with $a \sim y$ and $b \sim z$. Since y, z are both in A, which is connected, $y \sim z$. Since $\sim$ is symmetric and transitive, $a \sim b$. Therefore, there is a connected subspace $(C, \mathcal{T}_C)$ of $(S, \mathcal{T})$ with $a, b \in C$. But then $a \in U \cap C$, $b \in V \cap C$, and $(U \cap C) \cap (V \cap C) = \emptyset$, a contradiction. So, $(B, \mathcal{T}_B)$ is connected. Let $a \in A$. We will now show that $B = [a]$. First, suppose that $x \in B$. Then there is $y \in A$ with $x \sim y$. Since $a \in A$, $y \sim a$. Since $\sim$ is transitive, $x \sim a$. So, $x \in [a]$. Since $x \in B$ was arbitrary, $B \subseteq [a]$. Now, let $x \in [a]$. Then $x \sim a$. So, the statement $\exists y \in A(x \sim y)$ is true (let $y = a$). Therefore, $x \in B$. Since $x \in [a]$ was arbitrary, $[a] \subseteq B$. Since $B \subseteq [a]$ and $[a] \subseteq B$, we have $B = [a]$. □

(iii) Let $(A, \mathcal{T}_A)$ be a connected subspace of $(S, \mathcal{T})$ that is both open and closed in S and let $a \in A$. By (ii), A is contained in some component B. If $A \neq B$, then $\big((A \cap B), ((S \setminus A) \cap B)\big)$ would be a disconnection of A. Therefore, $A = B$. □

(iv) Let A be a component of S and assume toward contradiction that A is not closed in S. By Problem 10, $\overline{A}$ is a connected subspace of S, and since A is not closed, $A \neq \overline{A}$. Let $a \in A$ and $b \in \overline{A} \setminus A$. Then $a \sim b$, and so b is in the same component as a. Since $b \notin A$, this contradicts our assumption that A is a component. Therefore, A is closed in S. □

LEVEL 5

12. Let $(S, \mathcal{T})$ be a compact T_2-space. Prove that $\mathcal{T}$ has a basis $\mathcal{B}$ such that each set in $\mathcal{B}$ is both open and closed if and only if $(S, \mathcal{T})$ is totally disconnected.

Proof: Let $(S, \mathcal{T})$ be a T_2-space. If $\mathcal{T}$ has a basis $\mathcal{B}$ such that each set in $\mathcal{B}$ is both open and closed, then by Problem 8 above, $(S, \mathcal{T})$ is totally disconnected.

Conversely, assume that $(S, \mathcal{T})$ is totally disconnected. We will show that the collection of clopen sets in S is a basis for S. Let $x \in S$ and let U be an open set in S containing x. If $U = S$, then S is a clopen set containing x with $S \subseteq U$. So, assume $U \neq S$. Since U is open in S, $S \setminus U$ is closed in S. Since S is compact, $S \setminus U$ is also compact. Since $(S, \mathcal{T})$ is totally disconnected, for each $y \in S \setminus U$, there is a clopen set V_y with $y \in V_y$ and $x \notin V_y$. The collection $\mathcal{C} = \{V_y \mid y \in S \setminus U\}$ is an open covering of $S \setminus U$ by open sets in S. By the compactness of $(S, \mathcal{T})$, there is a finite subcover $\mathcal{D}$ of $\mathcal{C}$. Let $V = \bigcup \mathcal{D}$. Since a union of open sets is open, V is open in S. Since a finite union of closed sets is closed, V is also closed in S. It follows that $S \setminus V$ satisfies $x \in S \setminus V \subseteq U$, as desired. □

13. Prove that an arbitrary product of connected spaces is connected.

Proof: Let K be an index set, let $(S_k, \mathcal{T}_k)$ be connected for each $k \in K$, and assume toward contradiction that $\prod S_k$ is disconnected. By Problem 1 above, there is a surjective continuous function $f\colon \prod S_k \to \{0,1\}$ (here $\{0,1\}$ is given the discrete topology).

Let $\mathbf{x} \in \prod S_k$ and for some $j \in K$, define $f_j\colon S_j \to \prod S_k$ by $f_j(s) = (z_k)$, where $z_k = x_k$ for $k \neq j$ and $z_j = s$. Then f_j is continuous, and so, the function $g_j\colon S_j \to \{0,1\}$ defined by $g_j = f \circ f_j$ is continuous. So, by Problem 1 above, g_j is constant. Since $g_j(x_j) = f\left(f_j(x_j)\right) = f(\mathbf{x})$, it follows that $f\left(f_j(s)\right) = g_j(s) = f(\mathbf{x})$ for all $s \in S_j$. So, $f(\mathbf{s}) = f(\mathbf{x})$ whenever $\mathbf{s}$ agrees with $\mathbf{x}$ everywhere except possibly at s_j (f is constant on the jth "slice" of the product).

By repeating this process with another index, say $t \neq j$, we see that $f(\mathbf{s}) = f(\mathbf{x})$ whenever $\mathbf{s}$ agrees with $\mathbf{x}$ everywhere except possibly at s_t. It then follows that $f(\mathbf{s}) = f(\mathbf{x})$ whenever $\mathbf{s}$ agrees with $\mathbf{x}$ everywhere except possibly at s_j and s_t.

We can repeat this process finitely many times to see that $f(\mathbf{s}) = f(\mathbf{x})$ whenever $\mathbf{s}$ agrees with $\mathbf{x}$ everywhere except possibly at finitely many coordinates. The set of all $\mathbf{s}$ of this type is dense in $\prod S_k$ and so, f is constant, contradicting our assumption that f was surjective. □

Problem Set 15

LEVEL 1

1. Determine if each of the following subsets of $\mathbb{R}^2$ is a subspace of $\mathbb{R}^2$:
   (i) $A = \{(x, y) \mid x + y = 0\}$
   (ii) $B = \{(x, y) \mid xy = 0\}$
   (iii) $C = \{(x, y) \mid 2x = 3y\}$
   (iv) $D = \{(x, y) \mid x \in \mathbb{Q}\}$

Solutions:

(i) Since $0 + 0 = 0$, $(0, 0) \in A$.

Let $(x, y), (z, w) \in A$. Then $x + y = 0$ and $z + w = 0$.Therefore,

$$(x + z) + (y + w) = (x + y) + (z + w) = 0 + 0 = 0.$$

So, $(x, y) + (z, w) = (x + z, y + w) \in A$.

Let $(x, y) \in A$ and $k \in \mathbb{R}$. Then $x + y = 0$. So, $kx + ky = k(x + y) = k \cdot 0 = 0$ (by part (iii) of Problem 4 below).

So, $k(x, y) = (kx, ky) \in A$.

By Theorem 15.3, A is a subspace of $\mathbb{R}^2$.

(ii) Since $0 \cdot 1 = 0$, we have $(0, 1) \in B$. Since $1 \cdot 0 = 0$, we have $(1, 0) \in B$. Adding these two vectors gives us $(1, 0) + (0, 1) = (1, 1)$. However, $1 \cdot 1 = 1 \neq 0$, and so, $(1, 1) \notin B$. So, B is not closed under addition. Therefore, B is **not** a subspace of $\mathbb{R}^2$.

(iii) Since $2 \cdot 0 = 0$ and $3 \cdot 0 = 0$, $2 \cdot 0 = 3 \cdot 0$. Therefore, $(0, 0) \in C$.

Let $(x, y), (z, w) \in C$. Then $2x = 3y$ and $2z = 3w$.Therefore,

$$2(x + z) = 2x + 2z = 3y + 3w = 3(y + w).$$

So, $(x, y) + (z, w) = (x + z, y + w) \in C$.

Let $(x, y) \in C$ and $k \in \mathbb{R}$. Then $2x = 3y$. So, $2(kx) = k(2x) = k(3y) = 3(ky)$.

So, $k(x, y) = (kx, ky) \in C$.

By Theorem 8.1, A is a subspace of $\mathbb{R}^2$.

(iv) Since $1 \in \mathbb{Q}$, $(1, 0) \in D$. Now, $\sqrt{2}(1, 0) = (\sqrt{2}, 0) \notin D$ because $\sqrt{2} \notin \mathbb{Q}$. So, D is not closed under scalar multiplication. Therefore, D is **not** a subspace of $\mathbb{R}^2$.

2. Let X be a nonempty subset of a Banach space V. Prove that X is bounded if and only if there is a real number M such that for all $x \in X$, $|x| \leq M$, where $|x|$ is the norm of x in V.

Proof: First assume that X is bounded. Then there is $L \in \mathbb{R}^+$ so that for all $x, y \in X$, $|x - y| \leq L$. Let $y \in X$ be any element of X. Then for all $x \in X$, we have

$$|x| = |(x - y) + y| \leq |x - y| + |y| \leq L + |y|.$$

So, if we let $M = L + |y|$. Then for all $x \in X$, $|x| \leq M$.

Conversely, suppose that there is $M \in \mathbb{R}$ such that $|x| \leq M$ for all $x \in X$. Then for all $x, y \in X$, we have

$$|x - y| \leq |x| + |y| \leq 2M.$$

So X is bounded. □

LEVEL 2

3. Let $\mathbb{F}$ be a field. Prove that $\mathbb{F}^n$ is a vector space over $\mathbb{F}$.

Proof: We first prove that $(\mathbb{F}^n, +)$ is a commutative group.

(Closure) Let $(a_1, a_2, \dots, a_n), (b_1, b_2, \dots, b_n) \in \mathbb{F}^n$. Then $a_1, a_2, \dots, a_n, b_1, b_2, \dots, b_n \in \mathbb{F}$. By definition, $(a_1, a_2, \dots, a_n) + (b_1, b_2, \dots, b_n) = (a_1 + b_1, a_2 + b_2, \dots, a_n + b_n)$. Since $\mathbb{F}$ is closed under addition, $a_1 + b_1, a_2 + b_2, \dots, a_n + b_n \in \mathbb{F}$. Therefore,$(a_1, a_2, \dots, a_n) + (b_1, b_2, \dots, b_n) \in \mathbb{F}^n$.

(Associativity) Let $(a_1, a_2, \dots, a_n), (b_1, b_2, \dots, b_n), (c_1, c_2, \dots, c_n) \in \mathbb{F}^n$. Since addition is associative in $\mathbb{F}$, we have

$$\begin{aligned}
[(a_1, a_2, \dots, a_n) + (b_1, b_2, \dots, b_n)] + (c_1, c_2, \dots, c_n) &= (a_1 + b_1, a_2 + b_2, \dots, a_n + b_n) + (c_1, c_2, \dots, c_n) \\
&= \big((a_1 + b_1) + c_1, (a_2 + b_2) + c_2, \dots, (a_n + b_n) + c_n\big) \\
&= (a_1 + (b_1 + c_1), a_2 + (b_2 + c_2), \dots, a_n + (b_n + c_n)) \\
&= (a_1, a_2, \dots, a_n) + (b_1 + c_1, b_2 + c_2, \dots, b_n + c_n) \\
&= (a_1, a_2, \dots, a_n) + [(b_1, b_2, \dots, b_n) + (c_1, c_2, \dots, c_n)].
\end{aligned}$$

(Commutativity) Let $(a_1, a_2, \dots, a_n), (b_1, b_2, \dots, b_n) \in \mathbb{F}^n$. Since addition is commutative in $\mathbb{R}$, we have

$$\begin{aligned}
(a_1, a_2, \dots, a_n) + (b_1, b_2, \dots, b_n) = (a_1 + b_1, a_2 + b_2, \dots, a_n + b_n) &= (b_1 + a_1, b_2 + a_2, \dots, b_n + a_n) \\
&= (b_1, b_2, \dots, b_n) + (a_1, a_2, \dots, a_n).
\end{aligned}$$

(Identity) We show that $(0, 0, \dots, 0)$ is an additive identity for $\mathbb{F}^n$. Let $(a_1, a_2, \dots, a_n) \in \mathbb{F}^n$. Since 0 is an additive identity for $\mathbb{R}$, we have

$$(0, 0, \dots, 0) + (a_1, a_2, \dots, a_n) = (0 + a_1, 0 + a_2, \dots, 0 + a_n) = (a_1, a_2, \dots, a_n).$$

$$(a_1, a_2, \dots, a_n) + (0, 0, \dots, 0) = (a_1 + 0, a_2 + 0, \dots, a_n + 0) = (a_1, a_2, \dots, a_n).$$

(Inverse) Let $(a_1, a_2, \dots, a_n) \in \mathbb{F}^n$. Then $a_1, a_2, \dots, a_n \in \mathbb{F}$. Since $\mathbb{F}$ has the additive inverse property, $-a_1, -a_2, \dots, -a_n \in \mathbb{F}$. So, $(-a_1, -a_2, \dots, -a_n) \in \mathbb{F}^n$ and

$$(a_1, a_2, \dots, a_n) + (-a_1, -a_2, \dots, -a_n) = (a_1 - a_1, a_2 - a_2, \dots, a_n - a_n) = (0, 0, \dots, 0).$$

$$(-a_1, -a_2, \dots, -a_n) + (a_1, a_2, \dots, a_n) = (-a_1 + a_1, -a_2 + a_2, \dots, -a_n + a_n) = (0, 0, \dots, 0).$$

Now, let's prove that $\mathbb{F}^n$ has the remaining vector space properties.

(Closure under scalar multiplication) Let $k \in \mathbb{F}$ and let $(a_1, a_2, \ldots, a_n) \in \mathbb{F}^n$. Then $a_1, a_2, \ldots, a_n \in \mathbb{F}$. By definition, $k(a_1, a_2, \ldots, a_n) = (ka_1, ka_2, \ldots, ka_n)$. Since $\mathbb{F}$ is closed under multiplication, $ka_1, ka_2, \ldots, ka_n \in \mathbb{F}$. Therefore, $k(a_1, a_2, \ldots, a_n) \in \mathbb{F}^n$.

(Scalar multiplication identity) Let 1 be the multiplicative identity of $\mathbb{F}$ and let $(a_1, a_2, \ldots, a_n) \in \mathbb{F}^n$. Then $1(a_1, a_2, \ldots, a_n) = (1a_1, 1a_2, \ldots, 1a_n) = (a_1, a_2, \ldots, a_n)$.

(Associativity of scalar multiplication) Let $j, k \in \mathbb{F}$ and $(a_1, a_2, \ldots, a_n) \in \mathbb{F}^n$. Then since multiplication is associative in $\mathbb{F}$, we have

$$(jk)(a_1, a_2, \ldots, a_n) = \big((jk)a_1, (jk)a_2, \ldots, (jk)a_n\big) = (j(ka_1), j(ka_2), \ldots, j(ka_n))$$
$$= j(ka_1, ka_2, \ldots, ka_n) = j\big(k(a_1, a_2, \ldots, a_n)\big).$$

(Distributivity of 1 scalar over 2 vectors) Let $k \in \mathbb{F}$ and $(a_1, a_2, \ldots, a_n), (b_1, b_2, \ldots, b_n) \in \mathbb{F}^n$. Since multiplication is distributive over addition in $\mathbb{F}$, we have

$$k\big((a_1, a_2, \ldots, a_n) + (b_1, b_2, \ldots, b_n)\big) = k\big((a_1 + b_1, a_2 + b_2, \ldots, a_n + b_n)\big)$$
$$= \big(k(a_1 + b_1), k(a_2 + b_2), \ldots, k(a_n + b_n)\big) = \big((ka_1 + kb_1), (ka_2 + kb_2), \ldots, (ka_n + kb_n)\big)$$
$$= (ka_1, ka_2, \ldots, ka_n) + (kb_1, kb_2, \ldots, kb_n) = k(a_1, a_2, \ldots, a_n) + k(b_1, b_2, \ldots, b_n).$$

(Distributivity of 2 scalars over 1 vector) Let $j, k \in \mathbb{F}$ and $(a_1, a_2, \ldots, a_n) \in \mathbb{F}^n$. Since multiplication is distributive over addition in $\mathbb{F}$, we have

$$(j + k)(a_1, a_2, \ldots, a_n) = \big((j + k)a_1, (j + k)a_2, \ldots, (j + k)a_n\big)$$
$$= (ja_1 + ka_1, ja_2 + ka_2, \ldots, ja_n + ka_n) = (ja_1, ja_2, \ldots, ja_n) + (ka_1, ka_2, \ldots, ka_n)$$
$$= j(a_1, a_2, \ldots, a_n) + k(a_1, a_2, \ldots, a_n).$$

4. Let V be a vector space over $\mathbb{F}$. Prove each of the following:
   (i) For every $v \in V$, $-(-v) = v$.
   (ii) For every $v \in V$, $0v = 0$.
   (iii) For every $k \in \mathbb{F}$, $k \cdot 0 = 0$.
   (iv) For every $v \in V$, $-1v = -v$.

Proofs:

(i) Since $-v$ is the additive inverse of v, we have $v + (-v) = -v + v = 0$. But this equation also says that v is the additive inverse of $-v$. So, $-(-v) = v$. □

(ii) Let $v \in V$. Then $0v = (0 + 0)v = 0v + 0v$. So, we have

$$0 = -0v + 0v = -0v + (0v + 0v) = (-0v + 0v) + 0v = 0 + 0v = 0v.$$ □

(iii) Let $k \in \mathbb{F}$. Then $k \cdot 0 = k(0 + 0) = k \cdot 0 + k \cdot 0$. So, we have

$$0 = -k \cdot 0 + k \cdot 0 = -k \cdot 0 + (k \cdot 0 + k \cdot 0) = (-k \cdot 0 + k \cdot 0) + k \cdot 0 = 0 + k \cdot 0 = k \cdot 0.$$ □

(iv) Let $v \in V$. Then we have $v + (-1v) = 1v + (-1v) = (1 + (-1))v = 0v = 0$ by (ii) and we have $-1v + v = -1v + 1v = (-1 + 1)v = 0v = 0$ again by (ii). So, $-1v = -v$. □

LEVEL 3

5. Let V be a vector space over a field $\mathbb{F}$ and let $\boldsymbol{X}$ be a set of subspaces of V. Prove that $\bigcap \boldsymbol{X}$ is a subspace of V.

Proof: Let V be a vector space over a field $\mathbb{F}$ and let $\boldsymbol{X}$ a set of subspaces of V. For each $U \in \boldsymbol{X}$, $0 \in U$ because $U \leq V$. So, $0 \in \bigcap \boldsymbol{X}$. Let $v, w \in \bigcap \boldsymbol{X}$. For each $U \in \boldsymbol{X}$, $v, w \in U$, and so, $v + w \in U$ because $U \leq V$. Therefore, $v + w \in \bigcap \boldsymbol{X}$. Let $v \in \bigcap \boldsymbol{X}$ and $k \in \mathbb{F}$. For each $U \in \boldsymbol{X}$, $v \in U$, and so, $kv \in U$ because $U \leq V$. Therefore, $kv \in \bigcap \boldsymbol{X}$. By Theorem 15.3, $\bigcap \boldsymbol{X} \leq V$. □

6. Let $\mathcal{A} = \{f_n : [0,1] \to \mathbb{R} \mid n \in \mathbb{Z}^+\}$, where for each $n \in \mathbb{Z}^+$, $f_n(x) = \begin{cases} nx & \text{if } 0 \leq x < \frac{1}{n}. \\ 1 & \text{if } \frac{1}{n} \leq x \leq 1. \end{cases}$

Use the definition of equicontinuity to prove that $\mathcal{A}$ is not equicontinuous.

Proof: Let $\epsilon = \frac{1}{2}$ and suppose toward contradiction that there is $\delta_1 > 0$ such that for all $x, y \in K$ and every $f \in \mathcal{A}$, $d(x, y) < \delta$ implies $|f(x) - f(y)| < \frac{1}{2}$. Let $\delta = \min\{\delta_1, 1\}$. Let $x = 0$ and $y = \frac{\delta}{2}$. Then $|x - y| = \left|0 - \frac{\delta}{2}\right| = \frac{\delta}{2} < \delta$. By the Archimedean Property of $\mathbb{R}$, we can choose $n \in \mathbb{N}$ so that $n > \frac{2}{\delta}$, or equivalently, $\frac{1}{n} < \frac{\delta}{2}$. Then $f_n\left(\frac{\delta}{2}\right) = 1$. So, $\left|f_n(0) - f_n\left(\frac{\delta}{2}\right)\right| = |0 - 1| = 1 \geq \frac{1}{2}$, a contradiction. So, $\mathcal{A}$ is not equicontinuous. □

LEVEL 4

7. Let U and W be subspaces of a vector space V. Determine necessary and sufficient conditions for $U \cup W$ to be a subspace of V.

Theorem: Let U and W be subspaces of a vector space V. Then $U \cup W$ is a subspace of V if and only if $U \subseteq W$ or $W \subseteq U$.

Proof: Let U and W be subspaces of a vector space V. If $U \subseteq W$, then $U \cup W = W$, and so, $U \cup W$ is a subspace of V. Similarly, if $W \subseteq U$, then $U \cup W = U$, and so, $U \cup W$ is a subspace of V.

Suppose that $U \nsubseteq W$ and $W \nsubseteq U$. Let $x \in U \setminus W$ and $y \in W \setminus U$. Suppose that $x + y \in U$. We have $-x \in U$ because U is a subspace of V. So, $y = (-x + x) + y = -x + (x + y) \in U$, contradicting $y \in W \setminus U$. So, $x + y \notin U$. A similar argument shows that $x + y \notin W$. So, $x + y \notin U \cup W$. It follows that $U \cup W$ is not closed under addition, and therefore, $U \cup W$ is **not** a subspace of V. □

Note: The conditional statement $p \to q$ can be read "q is necessary for p" or "p is sufficient for q." Furthermore, $p \leftrightarrow q$ can be read "p is necessary and sufficient for q" (as well as "q is necessary and sufficient for p."

So, when we are asked to determine necessary and sufficient conditions for a statement p to be true, we are being asked to find a statement q that is logically equivalent to the statement p.

Usually if we are being asked for necessary and sufficient conditions, the hope is that we will come up with an equivalent statement that is easier to understand and/or visualize than the given statement.

8. Give an example of vector spaces U and V with $U \subseteq V$ such that U is closed under scalar multiplication, but U is not a subspace of V.

Solution: Let $V = \mathbb{R}^2$ and $U = \{(x, y) \mid x = 0 \text{ or } y = 0 \text{ (or both)}\}$. Let $(x, y) \in U$ and $k \in \mathbb{R}$. Then $k(x, y) = (kx, ky)$. If $x = 0$, then $kx = 0$. If $y = 0$, then $ky = 0$. So, $k(x, y) \in U$. So, U is closed under scalar multiplication. Now, $(0, 1)$ and $(1, 0)$ are in U, but $(1, 1) = (0, 1) + (1, 0) \notin U$. So, $U \not\leq V$.

LEVEL 5

9. Prove that $\mathcal{B}(S, \mathbb{R})$ together with the norm defined by $\|f\| = \sup|f(x)|$ is a Banach space.

Proof: By part 5 of Example 15.4, $\mathcal{B}(S, \mathbb{R})$ is a vector space. We now check that the 3 properties for a norm are satisfied.

If $f = \mathbf{0}$, then $f(x) = 0$ for all $x \in S$, and so, $\|f\| = \sup|f(x)| = 0$. On the other hand, if $\|f\| = 0$, then $f(x) = 0$ for all $x \in X$, and so, $f = \mathbf{0}$. So, Property 1 holds.

Let $f, g \in \mathcal{B}(S, \mathbb{R})$. For each $x \in X$, we have $|f(x) + g(x)| \leq |f(x)| + |g(x)|$, by the triangle Inequality. By the definition of the norm, for each $x \in X$, $|f(x) + g(x)| \leq \|f\| + \|g\|$. Again, by the definition of the norm, $\|f + g\| \leq \|f\| + \|g\|$. So, Property 2 holds.

Let $f \in \mathcal{B}(S, \mathbb{R})$ and let $c \in \mathbb{R}$. For each $x \in X$, we have $|c \cdot f(x)| = |c| \cdot |f(x)|$. Therefore, we have $\|cf\| = \sup|(cf)(x)| = \sup(|c| \cdot |f(x)|) = |c| \cdot \sup|f(x)| = |c| \cdot \|f\|$. So, Property 3 holds.

We now show that $(\mathcal{B}(S, \mathbb{R}), d)$ is complete, where $d: \mathcal{B}(S, \mathbb{R}) \times \mathcal{B}(S, \mathbb{R}) \to \mathbb{R}$ is defined by $d(f, g) = \|f - g\|$. To see this, let (f_n) be a Cauchy sequence in $\mathcal{B}(S, \mathbb{R})$. Then for each $x \in S$, $(f_n(x))$ is a Cauchy sequence in $\mathbb{R}$. Since $\mathbb{R}$ is complete, for each $x \in S$, there is $f(x) \in \mathbb{R}$ such that $f_n(x) \to f(x)$.

We first show that $f \in \mathcal{B}(S, \mathbb{R})$. So, we must show that f is bounded. Since (f_n) is a Cauchy sequence, there is $K \in \mathbb{N}$ such that $m, n > K$ implies $\|f_m - f_n\| < 1$. Then for all $x \in S$, we have

$$|f(x)| = \left|\left(f(x) - f_{K+1}(x)\right) + f_{K+1}(x)\right| \leq |f(x) - f_{K+1}(x)| + |f_{K+1}(x)|.$$

For $n > K$, we have $|f_n(x) - f_{K+1}(x)| < 1$, and so, $|f(x) - f_{K+1}(x)| \leq 1$ because $f_n(x) \to f(x)$. Therefore, $\|f - f_{K+1}\| = \sup|f(x) - f_{K+1}(x)| \leq 1$. It follows that for all $x \in S$,

$$|f(x)| \leq |f(x) - f_{K+1}(x)| + |f_{K+1}(x)| \leq 1 + \|f_{K+1}\|.$$

Thus, f is bounded.

Finally, we show that $f_n \to f$ in the metric d. Let $\epsilon > 0$. Since (f_n) is a Cauchy sequence, there is $K \in \mathbb{N}$ such that $m, n > K$ implies $|f_m - f_n| < \frac{\epsilon}{2}$. Then for $n > K$, we have $|f(x) - f_n(x)| \leq \frac{\epsilon}{2}$. Therefore, $|f_n - f| < \frac{\epsilon}{2} < \epsilon$. So, $f_n \to f$. □

10. Prove that $\mathcal{C}(S, \mathbb{R})$ is a closed subspace of $\mathcal{B}(S, \mathbb{R})$.

Proof: Let $f \in \overline{\mathcal{C}(S, \mathbb{R})}$. Then there is a sequence (f_n) in $\mathcal{C}(S, \mathbb{R})$ such that $f_n \to f$. Since $\mathcal{B}(S, \mathbb{R})$ is complete, $f \in \mathcal{B}(S, \mathbb{R})$. So, we just need to show that f is continuous. To this end, let $x \in S$ and let $\epsilon > 0$. Since $f_n \to f$, there is $K \in \mathbb{N}$ such that $n > K$ implies $|f_n - f| < \frac{\epsilon}{3}$. In particular, $|f_{K+1} - f| < \frac{\epsilon}{3}$. Since f_{K+1} is continuous at x, there is $\delta > 0$ such that $d(x, y) < \delta$ implies $|f_{K+1}(x) - f_{K+1}(y)| < \frac{\epsilon}{3}$. Therefore, if $d(x, y) < \delta$, then

$$\begin{aligned}|f(x) - f(y)| &= |(f(x) - f_{K+1}(x)) + (f_{K+1}(x) - f_{K+1}(y)) + (f_{K+1}(y) - f(y))| \\ &\leq |f(x) - f_{K+1}(x)| + |f_{K+1}(x) - f_{K+1}(y)| + |f_{K+1}(y) - f(y)| \\ &< 2\sup|f(y) - f_{K+1}(y)| + \frac{\epsilon}{3} \leq 2 \cdot \frac{\epsilon}{3} + \frac{\epsilon}{3} = \epsilon.\end{aligned}$$

So, f is continuous at x. Since $x \in S$ was arbitrary, $f \in \mathcal{C}(S, \mathbb{R})$. Since $f \in \overline{\mathcal{C}(S, \mathbb{R})}$, it follows that $\overline{\mathcal{C}(S, \mathbb{R})} \subseteq \mathcal{C}(S, \mathbb{R})$, and so $\mathcal{C}(S, \mathbb{R})$ is closed in $\mathcal{B}(S, \mathbb{R})$. □

11. Find a sequence (f_n) in $\mathcal{C}([0,1], \mathbb{R})$ such that f_n converges pointwise to $f \in \mathcal{C}([0,1], \mathbb{R})$, but (f_n) does **not** converge uniformly to f.

Solution: For each $n \geq 2$, define $f_n \in \mathcal{C}([0,1], \mathbb{R})$ by $f_n(x) = \begin{cases} nx & \text{if } 0 \leq x < \frac{1}{n}. \\ 2 - nx & \text{if } \frac{1}{n} \leq x < \frac{2}{n}. \\ 0 & \text{if } \frac{2}{n} \leq x \leq 1. \end{cases}$

It is easy to see that this sequence converges to $\mathbf{0}$ pointwise.

However, (f_n) does not converge to $\mathbf{0}$ uniformly because for any $n \in \mathbb{N}$, we have

$$|f_n - f| = \sup|f_n(x) - f(x)| = \sup|f_n(x)| = 1.$$

LEVEL 1

1. Prove that every path connected space is connected.

Proof: Let $(S,\mathcal{T})$ be a topological space that is path connected and assume toward contradiction that $(S,\mathcal{T})$ is disconnected. Let (U,V) be a disconnection of S, let $x \in U$,and let $y \in V$. Since $(S,\mathcal{T})$ is path connected, there is a continuous function $f\colon [0,1] \to S$ such that $f(0) = x$ and $f(1) = y$. By the continuity of f, $f^{-1}[U]$ and $f^{-1}[V]$ are both open in $[0,1]$. Also, $f^{-1}[U] \cap f^{-1}[V] = \emptyset$, $f^{-1}[U] \cup f^{-1}[V] = S$, $0 \in f^{-1}[U]$ and $1 \in f^{-1}[V]$. Therefore, $(f^{-1}[U], f^{-1}[V])$ is a disconnection of $[0,1]$, contradicting Theorem 14.3. It follows that $(S,\mathcal{T})$ is connected. □

LEVEL 2

2. Let $(X,\mathcal{T})$ and $(Y,\mathcal{U})$ be topological spaces, let $h\colon X \to Y$ be continuous, let $f,g\colon [0,1] \to X$ be paths in X, and let $F\colon [0,1] \times [0,1] \to X$ be a path homotopy from f to g. Prove that $h \circ F\colon [0,1] \times [0,1] \to Y$ is a path homotopy from $h \circ f$ to $h \circ g$. (This is Theorem 16.8.)

Proof: Suppose that f and g have initial and terminal points a and b, respectively. Since $h \circ F$ is a composition of continuous functions, it is continuous. Let $s \in [0,1]$. Then we have the following:

$$(h \circ F)(s,0) = h\big(F(s,0)\big) = h\big(f(s)\big) = (h \circ f)(s)$$

$$(h \circ F)(s,1) = h\big(F(s,1)\big) = h\big(g(s)\big) = (h \circ g)(s)$$

$$(h \circ F)(0,t) = h\big(F(0,t)\big) = h(a)$$

$$(h \circ F)(1,t) = h\big(F(1,t)\big) = h(b)$$

So, $h \circ F\colon [0,1] \times [0,1] \to Y$ is a path homotopy from $h \circ f$ to $h \circ g$. □

3. Let $(X,\mathcal{T})$ and $(Y,\mathcal{U})$ be topological spaces, let $h\colon X \to Y$ be continuous, and let $f,g\colon [0,1] \to X$ be paths in X with $f(1) = g(0)$. Prove that $h \circ (f \star g) = (h \circ f) \star (h \circ g)$. (This is Theorem 16.9.)

Proof: Let $s \in [0,1]$. Then we have

$$\big((h \circ f) \star (h \circ g)\big)(s) = \begin{cases} (h \circ f)(2s) & \text{if } s \in \left[0,\frac{1}{2}\right] \\ (h \circ g)(2s-1) & \text{if } s \in \left[\frac{1}{2},1\right] \end{cases}$$

$$= \begin{cases} h\big(f(2s)\big) & \text{if } s \in \left[0,\frac{1}{2}\right] \\ h\big(g(2s-1)\big) & \text{if } s \in \left[\frac{1}{2},1\right] \end{cases} = \big(h \circ (f \star g)\big)(s).$$

Since $s \in [0,1]$ was arbitrary, $h \circ (f \star g) = (h \circ f) \star (h \circ g)$ □

LEVEL 3

4. Let $(X, \mathcal{T})$ be a topological space and define the relation $\sim$ on X by $a \sim b$ if and only if there is a path from a to b. Prove that $\sim$ is an equivalence relation and that each equivalence class is path connected.

Proof: Let $a \in X$ and define $f: [0,1] \to X$ by $f(x) = a$ for all $x \in X$. Since all constant functions are continuous, f is a path from a to itself. So, $a \sim a$, and therefore, $\sim$ is reflexive.

Let $a, b \in X$ with $a \sim b$. Then there is a path $f: [0,1] \to X$ from a to b. Define $g: [0,1] \to X$ by $g(x) = f(1-x)$. Then g is a composition of continuous functions, and therefore, g is continuous. Also, $g(0) = f(1-0) = f(1) = b$ and $g(1) = f(1-1) = f(0) = a$. So, g is a path form b to a. Thus, $b \sim a$, and so, $\sim$ is symmetric.

Let $a, b, c \in S$ with $a \sim b$ and $b \sim c$. Then there are path $f: [0,1] \to X$ from a to b and $g: [0,1] \to X$ from b to c. Define $h: [0,1] \to X$ by $h(x) = \begin{cases} f(2x) & \text{if } x \in \left[0, \frac{1}{2}\right]. \\ g(2x-1) & \text{if } x \in \left[\frac{1}{2}, 1\right]. \end{cases}$ Since $f\left(2 \cdot \frac{1}{2}\right) = f(1) = b$ and $g\left(2 \cdot \frac{1}{2} - 1\right) = g(0) = b$, by the Pasting Theorem, h is continuous. Aslo, $h(0) = f(2 \cdot 0) = f(0) = a$ and $h(1) = h(2 \cdot 1 - 1) = h(1) = c$. So, h is a path from a to c. Thus, $a \sim c$, and therefore, $\sim$ is transitive.

Since $\sim$ is reflexive, symmetric, and transitive, $\sim$ is an equivalence relation.

Now, let $b, c \in [a]$. Then $b \sim a$ and $c \sim a$. Since $\sim$ is symmetric, $a \sim c$. Since $\sim$ is transitive, $b \sim c$. So, there is a path from a to c. Since $b, c \in [a]$ were arbitrary, we see that $[a]$ is path connected. □

5. Prove the Pasting Theorem (Theorem 16.2).

Proof: Let $(X, \mathcal{T})$ and $(Y, \mathcal{U})$ be topological spaces, let A, B be closed subsets of X such that $X = A \cup B$, let $f: A \to Y$ and $g: B \to Y$ be continuous functions, and suppose that for all $x \in A \cap B$, $f(x) = g(x)$. Let $h: X \to Y$ be defined by $h(x) = \begin{cases} f(x) & \text{if } x \in A. \\ g(x) & \text{if } x \in B. \end{cases}$

To show that h is continuous, by Problem 1 in Problem Set 13, it suffices to show that for any closed set C in Y, $h^{-1}[C]$ is closed in X. So, let C be closed in Y.

We first show that $h^{-1}[C] = f^{-1}[C] \cup g^{-1}[C]$. To see this, let $x \in h^{-1}[C]$. Then $h(x) \in C$. Also, $x \in X = A \cup B$. So, $x \in A$ or $x \in B$. Without loss of generality, assume that $x \in A$. Then $h(x) = f(x)$. So, $f(x) \in C$, and therefore, $x \in f^{-1}[C]$. So, $x \in f^{-1}[C]$ or $x \in g^{-1}[C]$. Thus, $x \in f^{-1}[C] \cup g^{-1}[C]$. Since $x \in h^{-1}[C]$ was arbitrary, $h^{-1}[C] \subseteq f^{-1}[C] \cup g^{-1}[C]$.

Now, let $x \in f^{-1}[C] \cup g^{-1}[C]$. Then $x \in f^{-1}[C]$ or $g^{-1}[C]$. Without loss of generality, assume that $x \in f^{-1}[C]$. Then $f(x) \in C$. Also, $x \in A$. So, $h(x) = f(x)$. So, $h(x) \in C$, and therefore, $x \in h^{-1}[C]$. Since $x \in f^{-1}[C] \cup g^{-1}[C]$ was arbitrary, $f^{-1}[C] \cup g^{-1}[C] \subseteq h^{-1}[C]$.

It follows that $h^{-1}[C] \subseteq f^{-1}[C] \cup g^{-1}[C]$.

Now, by the continuity of f and g, $f^{-1}[C]$ and $g^{-1}[C]$ are both closed in X. Therefore, their union is closed in X, and so, $h^{-1}[C]$ is closed in X, as desire. □

LEVEL 4

6. Let $(Y, \mathcal{U})$ be a simply connected topological space and let $f, g: [0,1] \to Y$ be paths in Y having the same initial and terminal points. Prove that $f \simeq_p g$.

Proof: Let $f, g: [0,1] \to Y$ be two paths from a to b. Then $f \star \overline{g}: [0,1] \to Y$ is a well-defined loop at a. Since $(Y, \mathcal{U})$ is simply connected, the loop $f \star \overline{g}$ is path homotopic to the constant loop c_a. We have

$$[f \star \overline{g}] \,\overline{\star}\, [g] = [c_a] \,\overline{\star}\, [g]$$

So, $[f \star \overline{g}] = [c_a]$, and therefore, $[f] \,\overline{\star}\, [\overline{g}] = [c_a]$. Thus, $[f] = [c_a] \,\overline{\star}\, [g] = [g]$. So, $f \simeq_p g$. □

LEVEL 5

7. Prove that topologically equivalent topological spaces have isomorphic fundamental groups.

Proof: $(X, \mathcal{T})$ and $(Y, \mathcal{U})$ be topologically equivalent and let $\phi: X \to Y$ be a homeomorphism. Define $\Phi: \pi_1(X, a) \to \pi_1(Y, b)$ by $\Phi([f]) = [\phi \circ f]$.

Φ is well-defined. To see this, let F be a path homotopy between the paths f and g. Then $\phi \circ F$ is a path homotopy between $\phi \circ f$ and $\phi \circ g$.

Φ is a homomorphism. To see this, observe that

$$\begin{aligned}\Phi([f] \,\overline{\star}\, [g]) &= \Phi([f \star g]) = [\phi \circ (f \star g)] = [(\phi \circ f) \star (\phi \circ g)] \\ &= [\phi \circ f] \,\overline{\star}\, [\phi \circ g] = \Phi([f]) \,\overline{\star}\, \Phi([g]).\end{aligned}$$

We show that Φ is bijective by producing an inverse $\Psi: \pi_1(Y, b) \to \pi_1(X, a)$. We do this by letting $\Psi([f]) = [\phi^{-1} \circ f]$. We have $\Phi(\Psi([f])) = \Phi([\phi^{-1} \circ f]) = [\phi \circ (\phi^{-1} \circ f)] = [f]$ and we also have $\Psi(\Phi([f])) = \Psi([\phi \circ f]) = [\phi^{-1} \circ (\phi \circ f)] = [f]$.

It follows that $\pi_1(X, a) \cong \pi_1(Y, b)$. □

About the Author

Dr. Steve Warner, a New York native, earned his Ph.D. at Rutgers University in Pure Mathematics in May 2001. While a graduate student, Dr. Warner won the TA Teaching Excellence Award.

After Rutgers, Dr. Warner joined the Penn State Mathematics Department as an Assistant Professor and in September 2002, he returned to New York to accept an Assistant Professor position at Hofstra University. By September 2007, Dr. Warner had received tenure and was promoted to Associate Professor. He has taught undergraduate and graduate courses in Precalculus, Calculus, Linear Algebra, Differential Equations, Mathematical Logic, Set Theory, and Abstract Algebra.

From 2003 – 2008, Dr. Warner participated in a five-year NSF grant, "The MSTP Project," to study and improve mathematics and science curriculum in poorly performing junior high schools. He also published several articles in scholarly journals, specifically on Mathematical Logic.

Dr. Warner has nearly two decades of experience in general math tutoring and tutoring for standardized tests such as the SAT, ACT, GRE, GMAT, and AP Calculus exams. He has tutored students both individually and in group settings.

In February 2010 Dr. Warner released his first SAT prep book "The 32 Most Effective SAT Math Strategies," and in 2012 founded Get 800 Test Prep. Since then Dr. Warner has written books for the SAT, ACT, SAT Math Subject Tests, AP Calculus exams, and GRE. In 2018 Dr. Warner released his first pure math book called "Pure Mathematics for Beginners." Since then he has released several more books, each one addressing a specific subject in pure mathematics.

Dr. Steve Warner can be reached at

steve@SATPrepGet800.com

BOOKS BY DR. STEVE WARNER

PHYSICS MASTERY
for the Advanced High School Student
Complete Physics Review
with 400 SAT and AP Physics Questions

320
GRE MATH
PROBLEMS
arranged by Topic
and Difficulty Level
By Dr. Steve Warner

Second Edition
320
AP CALCULUS
AB PROBLEMS
arranged by Topic
and Difficulty Level
By Dr. Steve Warner

GET 800
Second Edition
320
AP CALCULUS
BC PROBLEMS
arranged by Topic
and Difficulty Level
By Dr. Steve Warner

320
SAT
CHEMISTRY
SUBJECT TEST
PROBLEMS
arranged by Topic
and Difficulty Level
Christopher Bozza
Dr. Steve Warner

NEW
SHSAT
Verbal Prep Book
To Improve Your Score
In Two Months

1000 Logic and Reasoning Questions for
GIFTED AND TALENTED
Elementary School Students

Vocabulary Builder:
The Complete
Vocabulary Word
Workshop
for Building
English
Vocabulary
BONUS
INSIDE!

Algebra Handbook

555 Geometry
Problems
for High School Students

555 Math IQ
Questions
for Middle School Students

555 Advanced
Math Problems

www.ingramcontent.com/pod-product-compliance
Lightning Source LLC
Chambersburg PA
CBHW081818200326
41597CB00023B/4302
* 9 7 8 1 9 5 1 6 1 9 0 3 9 *